FUNDAMENTALS OF INDUSTRIAL CHEMISTRY

FUNDAMENTALS OF INDUSTRIAL CHEMISTRY

PHARMACEUTICALS, POLYMERS, AND BUSINESS

John A. Tyrell
University of North Carolina Wilmington

Published by John Wiley & Sons, Inc., Hoboken, New Jersey
Published simultaneously in Canada

Library of Congress Cataloging-in-Publication Data:

Tyrell, John A.
Fundamentals of industrial chemistry: pharmaceuticals, polymers, and business / John A. Tyrell, University of North Carolina, Wilmington.
pages cm
Includes bibliographical references and index.
ISBN 978-1-118-61756-4 (cloth)
1. Chemistry, Technical. 2. Chemistry, Technical—Study and teaching. I. Title.
TP145.T87 2014
661—dc23

2013051309

Printed in the United States of America

ISBN: 9781118617564

10 9 8 7 6 5 4 3 2 1

To my daughters, Christine Reichert, Catherine Backmeyer and Alicia Tyrell
and to the next generation: Sarah, Madeline, John, Anna and Abigail.

CONTENTS

CHAPTER 1

Introduction

The chemical industry is one of America's largest; a $750 billion dollar enterprise, it represents more than 13% of U.S. manufacturing exports, and directly employs more than 780,000 [1]. About 54% of all chemists work in manufacturing, 37% in academia, and the remainder are self-employed or in the government. Bachelor's degree chemists represent the group most employed in industry at 83% [2]. If you are a chemist, there is a likelihood that you work in industry (either in manufacturing, analytical services, or research services). If you don't work in industry, knowledge of industrial chemistry is still important. If you are in academics, you might be collaborating with industrial colleagues or preparing students for industrial careers. Many government chemists have jobs associated with the chemical industry and work closely with industrial colleagues. Even if you are not a chemist, you are surrounded by chemicals and chemical products and knowledge of the chemical industry is useful.

There is a misconception among some that research occurs in academia and little goes on in industry. This is not true at all. Chemical companies expend major time and money on research and employ many chemists in that endeavor. For example, BASF ($2.0B), Bayer ($1.3B), Dow ($1.7B), and Dupont ($1.7B) each year spend much more than $1 billion on research [3].

This text discusses how and why major chemicals are manufactured. Intertwined in these discussions are concepts such as separation techniques, cost, conversion, transport, byproduct formation, and other items critical to industrial chemistry. Many of the major chemicals are discussed. Also discussed are several different industries. Most of the largest volume organic chemicals that are produced are made as feedstocks for polymers. For this reason, polymer chemistry is given special attention. The text discusses many of the major industrial polymers including their synthesis and properties. A background in polymer science is also presented so that the reader becomes familiar with some important concepts such as glass transition,

Fundamentals of Industrial Chemistry: Pharmaceuticals, Polymers, and Business, First Edition. John A. Tyrell.

molecular weight, and additives. Many chemists work in the pharmaceutical industry and there is a discussion on this industry including some of the requirements such as GMP. Patent protection is critical for many industries. The importance of patents and patentability requirements are explained.

Some important inorganic chemicals, in approximate order of quantity produced, are: sulfuric acid (H_2SO_4), lime which is also known as calcium oxide (CaO), phosphoric acid (H_3PO_4), sodium hydroxide which is also known as caustic or caustic soda ($NaOH$), sodium chloride or salt ($NaCl$), sodium carbonate (Na_2CO_3), nitric acid (HNO_3), ammonium nitrate (NH_4NO_3) and titanium dioxide (TiO_2). Some important gases include nitrogen (N_2), oxygen (O_2), ammonia (NH_3), hydrogen (H_2), chlorine (Cl_2), and hydrochloric acid (HCl).

Some major organic compounds are:

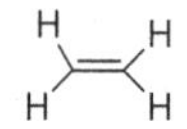

Ethylene 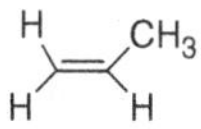   Propylene 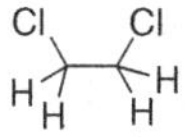   Ethylene dichloride (EDC) 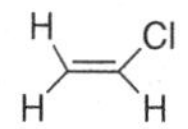   Vinyl chloride (or vinyl chloride monomer, VCM)

Benzene 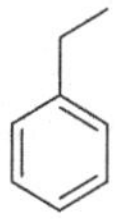   Ethylbenzene 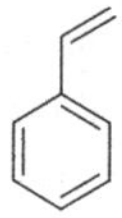   Styrene 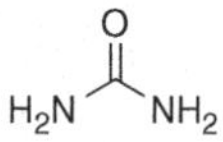   Urea 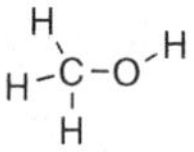   Methanol 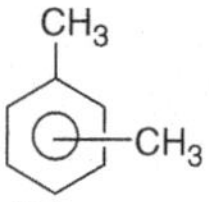   Xylene

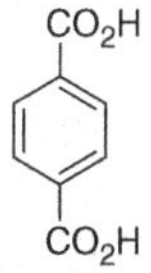

Terephthalic acid 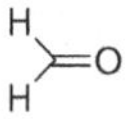   Formaldehyde   Ethylene oxide 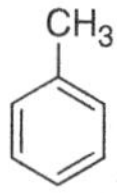   Toluene 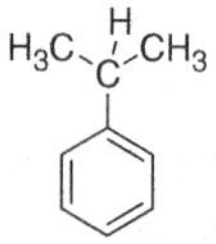   Cumene    Ethylene glycol

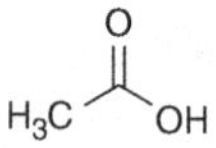

Acetic acid 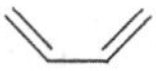   Butadiene 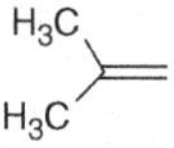   Isobutene = isobutylene 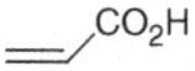   Acrylic acid 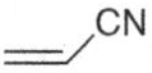   Acrylonitrile

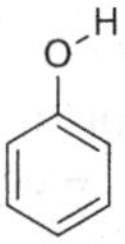

Phenol 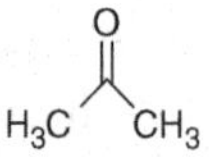   Acetone 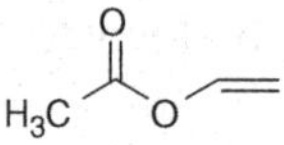   Vinyl acetate 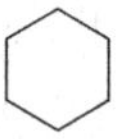   Cyclohexane 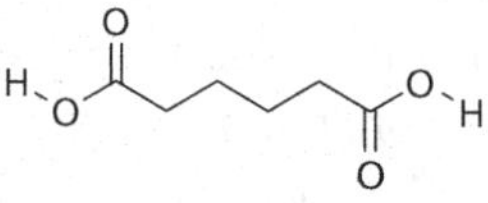   Adipic acid

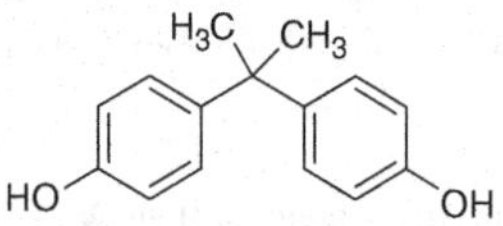

Bisphenol-A (BPA) 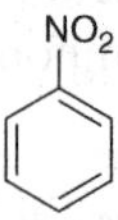   Nitrobenzene 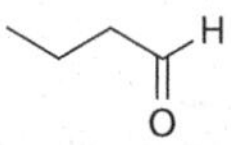   Butyraldehyde

Many of these chemicals are intertwined with each other. For example, chlorine is coproduced with sodium hydroxide and is reacted with ethylene to make ethylene dichloride, which in turn is used to make vinyl chloride. Many of the organic chemicals are produced to make polymers. For example, vinyl chloride is used to make polyvinyl chloride. The upcoming chapters will discuss these relationships and also the larger volume polymers.

The text is written for the student that would like to give their chemistry classes some perspective and perhaps learn something about chemical applications. The typical student will be a chemistry or chemical engineering student with at least a couple of years of classes. It is written for the upper-level undergraduate student or the first year graduate student. First or second year employees in the chemical industry will also benefit from the text.

The text assumes the reader has taken general and organic chemistry. Complete retention of everything from those courses is not assumed and when a concept is introduced, the background is given. The purpose of the text is to give the reader a general knowledge of several different aspects of industrial chemistry. It is not intended to make the reader an instant expert in a specific area. For example, reading the chapter on patents will give an appreciation of the major requirements for patentability, some key concepts and the importance of patents. It will also explain how to search patents and what can be learned from the patent literature. It will not make the reader ready to practice patent law nor be able to head the legal department. However, the reader will be better able to interact with patent attorneys and patent agents. Having a broad knowledge of several areas is also useful because over a career, the reader's work is likely to evolve and include increasing and broadened responsibilities.

REFERENCES

1. Melody Bomgardner, et al. Chemical and Engineering News 2013; 91(26):25–48.
2. Sophie Rovner. Chemical and Engineering News 2013; 91(38):10.
3. Alexander Tullo. Chemical and Engineering News 2011; 89(30):14.

CHAPTER 2

Inorganic Chemicals

2.1 SULFURIC ACID

Sulfuric acid (H_2SO_4) is a large volume chemical made from sulfur dioxide, which in turn can be made from elemental sulfur. Because sulfuric acid is of prime importance to the world's fertilizer and manufacturing industries, consumption of sulfuric acid has been regarded as one of the indexes of a nation's industrial development [1]. Sulfuric acid is the largest volume chemical produced in the world. Sulfur is oxidized to sulfur dioxide. Sulfur dioxide is then further oxidized to sulfur trioxide. Temperatures of 400 °C to 450 °C are typical and a vanadium catalyst such as vanadium pentoxide (V_2O_5) is commonly used [2, 3]. At a lower temperature, sulfur trioxide combines with water to form sulfuric acid. The following reactions give a synopsis of the chemistry.

$$S + O_2 \longrightarrow SO_2$$

$$2\ SO_2 + 2\ V_2O_5 \rightleftharpoons 2\ SO_3 + 2\ V_2O_4$$

$$O_2 + 2\ V_2O_4 \rightleftharpoons V_2O_5$$

$$SO_3 + H_2O \rightleftharpoons H_2SO_4$$

There are several sources of sulfur. Elemental sulfur is naturally occurring and can be mined by a process invented in the late 19th century by Herman Frasch. The Frasch process takes advantage of the relatively low melting point of sulfur at 115 °C. Superheated water at 168 °C is pumped through pipes inserted into a well and molten sulfur is pumped from the well [4].

Another source is pyrite. Pyrite is iron sulfide (FeS_2). Pyrite is also known as fool's gold because of its visual resemblance to the precious metal. With

Fundamentals of Industrial Chemistry: Pharmaceuticals, Polymers, and Business, First Edition. John A. Tyrell.

water and oxygen, pyrite can be converted to sulfuric acid. China is a leading miner of pyrite and extraction of sulfuric acid from pyrite is an important process in China. The pyrite is roasted to form sulfur dioxide which is then purified and converted to sulfuric acid [5].

A major source of sulfur is refinery and natural gas streams. This is done by the Claus process which was discovered more than 100 years ago and has been used by the natural gas and refinery industries for 50 years. In the Claus process, hydrogen sulfide from the gas stream is converted to elemental sulfur. Air is introduced into a furnace to oxidize about one-third of the hydrogen sulfide to sulfur dioxide. In the next stage, the reaction furnace, unconverted hydrogen sulfide reacts with the sulfur dioxide to form elemental sulfur. The Claus process generally produces an overall recovery of sulfur of 95–97%, but several modifications have been invented and sulfur recoveries of 99.9% are now achievable [6]. The chemistry is represented by the following reactions; the equilibrium to form elemental sulfur is favored at lower temperatures.

$$2\,H_2S + 3\,O_2 \longrightarrow 2\,SO_2 + 2\,H_2O$$

$$2\,H_2S + SO_2 \rightleftharpoons 3\,S + 2\,H_2O$$

The leading U.S. producers are the refining companies such as Valero Energy Corp., Exxon Mobil Corp., Conoco Phillips Co., Chevron Oil Co., and Shell Oil Co. In 2011, elemental sulfur and byproduct sulfuric acid were produced in the United States at 109 operations in 29 states and St. Croix with total shipments being valued at $1.6 billion [7]. The production took place at petroleum refineries, natural-gas processing plants, and coking plants. About 90% of consumed sulfur is in the form of sulfuric acid. Production of fertilizers is the major use of sulfuric acid, followed by petroleum refining and metal mining. U.S. production data for 2009 is estimated at 9.8 million metric tons of sulfur. If this number is converted to a sulfuric acid basis, it becomes 30 million metric tons.

$$9.8 \text{ million metric tons} \times \frac{\text{98 formula weight sulfuric acid}}{\text{32 formula weight sulfur}} = 30 \text{ million metric tons}$$

The population of the U.S. is around 300 million people, so this is equivalent to each and every person in the United States making 0.1 metric tons each year. A metric ton is 1,000 kg or 2,240 pounds (not to be confused with a U.S. ton which is 2,000 pounds) so this is equivalent to each person making

224 pounds of sulfuric acid every year. The actual amount used is about 20% higher because of imports.

Reagent sulfuric acid is 95–98% by weight sulfuric acid with the remainder being water. It is a clear, colorless, oily, corrosive liquid. Fuming sulfuric acid, also known as oleum, is sulfuric acid with dissolved sulfur trioxide. Oleum is sold at a variety of percentages that indicate the amount of sulfur trioxide. For example, 24% oleum is 24 g sulfur trioxide per 100 g total. Recognize that this is also true for other mass units. So 24% oleum is also 24 pounds sulfur trioxide per 100 pounds total or 24 tons sulfur trioxide per 100 tons total. This is also sometimes expressed as 105.4% H_2SO_4, because 100 g of 24% oleum can make 105.4 g H_2SO_4 when the oleum reacts with water. This is a theoretical exercise and generally should not actually be done because oleum reacts violently with water. It is worthwhile to review the arithmetic because such exercises commonly need to be done in industry. The calculation can be done by converting the grams of sulfur trioxide per 100g total to moles of sulfur trioxide, multiplying by 18 g per mole of water, and adding that number to 100. The calculation is demonstrated for 34% oleum, which can be called 107.65% H_2SO_4.

$$34\ \text{g SO}_3 \times \frac{1\ \text{mole SO}_3}{80\ \text{g}} \times \frac{1\ \text{mole H}_2\text{O}}{1\ \text{mole SO}_3} \times \frac{18\ \text{g}}{1\ \text{mole H}_2\text{O}} = 7.65\ \text{g H}_2\text{O}$$

Sulfuric acid has many uses but one of the largest uses is for fertilizer manufacture. Primary nutrients for plants are nitrogen, phosphorus and potassium. The major use of sulfuric acid in fertilizer manufacture is in the production of phosphoric acid, which in turn is used to produce phosphorus-containing fertilizers.

2.2 PHOSPHORIC ACID

A major use of sulfuric acid is in the manufacture of phosphoric acid (H_3PO_4), also known as orthophosphoric acid. The principle use of phosphoric acid is to make fertilizers. In the United States, phosphate rock ore is mined by six companies at 12 mines and upgraded to 28 million metric tons [8]. Most of the mining is done in Florida and North Carolina. China is the leading producer of phosphate rock with 72 million metric tons of the 191 million metric tons manufactured worldwide.

Phosphorus is an essential element for plant and animal nutrition and phosphate rock is the only significant resource of phosphorus [9]. Phosphate rock is converted into phosphoric acid, which in turn is converted into phosphorus fertilizers. Common examples of phosphorous-containing fertilizers are triple superphosphate, monoammonium phosphate, diammonium phosphate, and ammonium polyphosphate. The latter three examples also provide nitrogen, another primary nutrient. Triple superphosphate is calcium dihydrogen phosphate (CaH_2PO_4). Two processes, the dry process and the wet process, are used to convert phosphate rock to phosphoric acid. In the manufacture of phosphoric acid by the dry process, phosphate rock is reduced in an electric furnace, and the resulting yellow phosphorus is burnt into the oxide, P_4O_{10}, commonly referred to by its empirical formula, P_2O_5, and therefore called phosphorous pentoxide. The phosphorous pentoxide is hydrated to obtain phosphoric acid. This process has high energy costs. The wet process is the more commonly used. In the wet process, phosphate rock is decomposed with sulfuric acid, and the produced calcium sulfate is separated to obtain dilute phosphoric acid, which is then concentrated into high concentration acid [10]. Direct acidification with sulfuric acid is problematic because the outside surface of the rock reacts with the sulfuric acid resulting in a gypsum layer on the outside of the rock. Gypsum is calcium sulfate dihydrate and is used for plaster and drywall. This gypsum layer limits the reaction on the inside of the rock. To address this problem, the phosphate rock is first acidified with phosphoric acid to make calcium dihydrogen phosphate. Sulfuric acid is then added to make phosphoric acid [11].

$$2\ Ca_5F(PO_4)_3 + 12\ H_3PO_4 \longrightarrow 9\ Ca(H_2PO_4)_2 + CaF_2$$

$$9\ Ca(H_2PO_4)_2 + CaF_2 + 10\ H_2SO_4 + 20\ H_2O \longrightarrow 18\ H_3PO_4 + 2\ HF + 10\ CaSO_4 \cdot 2H_2O$$

This process requires that 12 of the 18 moles, or 67%, of phosphoric acid produced be recycled to treat more phosphate rock. There are other metals in the phosphate rock and upon acidification they become salts. The insoluble salts remain with the calcium sulfate precipitate but the soluble salts such as magnesium sulfate or iron sulfate remain in solution with the phosphoric acid. Generally, high purity is not required for fertilizer grade phosphoric acid. The crude phosphoric acid reacts, for example, with ammonia to make ammonium dihydrogen phosphate.

$$NH_3 + H_3PO_4 \longrightarrow (NH_4)H_2PO_4$$

If higher purity is required, such as for food applications, there are various processes to improve the purity of wet process phosphoric acid. Generally they involve a solvent which solubilizes the phosphoric acid, but not the impurities [12]. Alternatively, dry process phosphoric acid can be used. However, purification of wet process phosphoric acid is the dominant route because of the higher costs associated with the thermal process.

Phosphoric acid is a low melting solid and it is usually sold as an aqueous solution, such as 85% phosphoric acid, which freezes at about 21 °C. More dilute solutions are also sold and they have lower freezing points. For example, 75% phosphoric acid has a freezing point of −18 °C and can be used when heated storage is not available. Sometimes phosphoric acid is classified or sold based upon the amount of phosphorous pentoxide present. A solution of 85% phosphoric acid is 61.5% P_2O_5. This can be calculated by taking into account that it takes two moles of phosphoric acid to make one mole of phosphorus pentoxide and using the molecular weights.

$$85\text{ g } H_3PO_4 \times \frac{1\text{ mole } H_3PO_4}{98\text{ g}} \times \frac{1\text{ mole } P_2O_5}{2\text{ mole } H_3PO_4} \times \frac{141.9\text{ g}}{1\text{ mole } P_2O_5} = 61.5\%\ P_2O_5$$

Although fertilizer production is the major use and represents almost 90% of U.S. consumption [13], there are many other applications for phosphoric acid. For example, it can be used to remove scale from boilers, to treat metals for improved corrosion resistance, or as a dilute solution in foods as an acidulant and flavoring agent.

2.3 LIME

Lime is calcium oxide (CaO). At the end of 2011, there were 72 lime plants operating in the United States [14]. They manufactured 19 million metric tons. This compares with a world production of 330 million metric tons. China is the major world producer with 200 million metric tons. Major markets for lime are steelmaking, treatment of smokestack emissions, construction, water treatment, mining, and the paper industry. Lime is often used as a base. It is used to trap acids, such as in the treatment of smokestack emissions to remove sulfur dioxide and other acidic gases. It is used in agriculture to adjust the pH of the soil. It is used to adjust pH levels of industrial waste water. It is also used for drinking water to remove trace metals by precipitation or flocculation. The decrease in consumption seen in 2009 is due to the effect of the recession on such industries as steel and construction. Lime is manufactured by the calcining of calcium carbonate ($CaCO_3$) in high temperature kilns. At the kiln temperatures, calcium carbonate decomposes to calcium oxide and carbon dioxide. Calcium carbonate is naturally occurring and found in rocks, coral, and sea shells. A rock containing at least 50% by weight calcium carbonate is classified as limestone and limestone mining is the principle source for calcium oxide manufacture. Limestone is readily available in the earth's crust. Energy costs for the calcining, done at temperatures of 900–1200 °C, are a major factor in the production cost. When lime is treated with water, calcium hydroxide ($Ca(OH)_2$), also known as slaked lime or hydrated lime, is formed. Calcium hydroxide has low water solubility (0.16 g/100 g water at 25 °C) [15] and a suspension in water is called "milk of lime."

2.4 SODA ASH

Soda ash is sodium carbonate (Na_2CO_3) and is a major chemical with about 11 million metric tons being produced each year in the United States [16]. About half of the produced soda ash is used in glass manufacture. Chemical manufacture is the second major use followed by soaps and detergents. Soda ash is traditionally produced from trona ore in the USA, and most of the soda ash production in the United States comes from Wyoming, which has the largest known trona ore deposit in the world [17]. Trona(sodium sesquicarbonate dihydrate, $Na_2CO_3 \cdot NaHCO_3 \cdot 2H_2O$) is a complex salt of sodium carbonate and sodium bicarbonate ($NaHCO_3$). Trona is converted to soda ash by calcination, followed by dissolution, filtration to remove insolubles, and crystallization of the monohydrate. After isolation, the hydrate is heated to remove water.

2.5 TITANIUM DIOXIDE

Titanium dioxide (TiO_2), also called titania, has excellent light-scattering properties and because of this, has found wide use as a white pigment. U.S. production of titanium dioxide is estimated at about 300,000 metric tons per year with another 1.1 million metric tons being imported [18]. Titanium is the ninth most abundant element in the earth's crust. Of the 83 elements found in nature, 12 make up 99.7 percent of the Earth's crust. In decreasing order of natural abundance they are: oxygen, silicon, aluminum, iron, calcium, magnesium, sodium, potassium, titanium, hydrogen, phosphorus, and manganese [19]. The main titanium ores are ilmenite ($FeTiO_3$) and rutile (TiO_2). Titanium ore is converted to pure titanium dioxide, a white crystalline powder.

Titanium dioxide imparts whiteness and brightness to objects. It makes paints, plastics, paper and even food white in color. It also imparts opacity, making objects and coatings less transparent. The major use is in paints and coatings. It imparts whiteness in white paints but even color paints use titanium dioxide to give hiding power. Paints without titanium dioxide are more "see through." Other major uses include plastics and paper, but titanium dioxide has widespread application including cosmetics, sunblock, and toothpaste.

Titanium dioxide exists in two major crystal forms, rutile and anatase. Rutile is more closely packed and therefore denser than anatase. Compared with anatase, it has a higher refractive index and higher opacity. Because of its brilliant whiteness, excellent covering power, and resistance to color change, rutile is a valuable pigment for a broad range of applications in paints, plastics, inks, and paper. Anatase has a bluer undertone and is less abrasive than rutile. It is often preferred for applications in paper, ceramics, rubber, and fibers.

Titanium dioxide has higher opacity than other white pigments. Rutile is about 15% higher than anatase but more than double that of zinc sulfide, more than three times that of antimony oxide, and more than ten times that of calcium carbonate.

There are two major processes: the sulfate process and the chloride process. In the sulfate process, titanium-bearing ore, typically ilmenite, is treated with sulfuric acid. The titanium dissolves as titanyl sulfate and is then hydrolyzed to titanium dioxide which forms as a precipitate. Filtration, washing and calcination give titanium dioxide crystals of the correct pigment size [20]. Crystal structure (rutile or anatase), size, and distribution are determined by process conditions and additives [21]. The sulfate process was the dominant process in the 1970s and 1980s and continues to be used today, but it is a lengthy process. Today, the chloride process dominates because of advantages in waste generation, energy usage, and quality improvements. There has not been a sulfate process plant built in the United States since 1971.

In the sulfate process, titanium ore is digested with sulfuric acid to form titanium sulfate ($TiOSO_4$). The titanium sulfate is then hydrolyzed, filtered, washed, and calcined to give titanium dioxide.

In the chloride process, dried ore reacts with chlorine gas to produce titanium tetrachloride. The titanium tetrachloride is purified chemically and by distillation and then reacted with oxygen to form titanium dioxide and chlorine. The chlorine is recycled. Prior to the oxidation, a minor amount of aluminum chloride is added to promote rutile formation [22]. The rutile is preferred because it is more durable and has a higher refractive index. Untreated titanium dioxide is used, but more commonly a surface treatment is done to improve specific end-use property requirements such as ability to disperse or weather resistance.

The chlorine used for titanium dioxide production is prepared by electrolysis of sodium chloride. Like titanium ore, sodium chloride is mined.

2.6 SODIUM CHLORIDE AND CHLORALKALI

Sodium chloride (NaCl), also called salt or halite, is widely available in the United States. In the United States, 28 companies with a total of 60 plants produce about 44 million metric tons of the 260 million metric tons produced worldwide [23]. Salt is obtained by underground mining of deposits. Mining is done by traditional dry mining and by solution mining. Solution mining involves injection of water into bedded or domal salt and then recovery of the brine [24]. Salt is also recovered from seawater and other saline waters by solar evaporation. In the United States, solar salt is a minor method of production. Salt for highway deicing is the major use but this is followed closely by consumption by the chemical industry, where the major use is as a feedstock for chloralkali plants. Salt is also used in agriculture, food, and in water treatment.

Chloralkali is a term that is used for the coproduction of chlorine (Cl_2) and caustic. Caustic, sometimes called caustic soda, is sodium hydroxide (NaOH). They are coproduced from salt and water by the following reaction.

$$2\,NaCl + 2\,H_2O \longrightarrow 2\,NaOH + Cl_2 + H_2$$

The reaction is an electrochemical reaction. At the anode, chloride is oxidized to form chlorine; at the cathode, water is reduced to form hydrogen and hydroxide.

Anode	$2\,Cl^- \longrightarrow Cl_2 + 2\,e^-$
Cathode	$2\,H_2O + 2\,e^- \longrightarrow H_2 + 2\,OH^-$
Overall	$2\,Na^+ + 2\,Cl^- + 2\,H_2O \longrightarrow 2\,Na^+ + 2\,OH^- + Cl_2 + H_2$

The process consumes a large quantity of electricity and energy costs are an important consideration for profitability. From the balanced equation, two moles of sodium chloride produce 2 moles of sodium hydroxide and 1 mole of chlorine gas. Don't confuse the molar ratio with the mass ratio. By using atomic masses, we can say that 117 g (2 moles × 58.5 g/mole) sodium chloride produces 80 g (2 moles × 40 g/mole) sodium hydroxide and 71 g chlorine (1 moles × 71 g/mole). Therefore, the mass ratio of caustic to chlorine is 80 to 71. The mass ratio is true for all mass units. A production of 80 pounds of caustic coproduces 71 pounds of chlorine. A plant consuming 117 tons of salt coproduces 80 tons of sodium hydroxide and 71 tons of chlorine. One of these is not produced without coproducing the other. This can be a problem in the marketplace when there is a demand for caustic but not chlorine or when there is a demand for chlorine but not caustic.

There are three major processes used for chloralkali production. They are mercury, diaphragm, and membrane and differ in the type of cell used. In the mercury cell process, the cathode is a mercury film. This process was dominant in the twentieth century, but has been declining. For example, in the U.S., Olin has announced that they will convert their mercury cell capacity by the end of 2012. This will leave Ashta Chemicals in Ashtabula, Ohio and PPG Industries in Natrium, West Virginia as the only remaining U.S. mercury cell producers [25]. One major reason is that environmental considerations make the use of large quantities of mercury less desirable. The mercury process also consumes more electricity than the other two processes. In a diaphragm process, the cathode and anode are separated by a permeable diaphragm. The membrane process is the newest of the three processes and is considered the most environmentally friendly and energy efficient. As the mercury process is phased out, the membrane process is becoming the dominant process.

A membrane is a layer of material that selectively allows one component to permeate through. In the case of the chloralkali process, there are many requirements that need to be met in the selection of the membrane material. The membrane needs to be chemically stable to chlorine and hypochlorite anion even at the high process temperatures of about 100 °C. There needs to be a low resistance which gives a low voltage drop across the membrane. Selectivity is important. The membrane needs to selectively allow sodium ions to pass through. Because of the demanding requirements, chloralkali membranes commonly employ perfluorinated polymers having carboxylate or sulfonate groups. These can be prepared by copolymerizing tetrafluoroethylene, a perfluorovinyl ether, and a perfluorinated comonomer containing, for example, a sulfonyl fluoride group which is subsequently hydrolyzed to a sodium sulfonate group. The perfluorosulfonic acid polymers were invented by Dupont scientists and commercialized under the trade name Nafion®.

The anode and cathode are separated by an ion exchange membrane which allows sodium ions to pass through. Sodium ions pass through the membrane to the cathode where water is reduced to hydrogen and hydroxide. Therefore high purity sodium hydroxide is prepared in the cathode compartment. Chloride does not pass through the membrane and is oxidized to chlorine in the anode compartment.

The major use for chlorine is the chlorination of ethylene to make ethylene dichloride (this is always referred to as ethylene dichloride, but the more proper name is 1,2-dichlorethane) by addition of chlorine to the double bond. Ethylene dichloride, or EDC, is converted into vinyl chloride, which is then converted to polyvinyl chloride (PVC) polymer. There are many uses for PVC, but much of the demand is in the building and construction industry for items such as plumbing, vinyl siding, and window frames. Therefore the robustness of the construction business has a major effect on PVC demand and, in turn, chlorine demand. Other uses for chlorine include water purification, titanium dioxide production, phosgene production and organic intermediates. In water purification, chlorine can be directly added to water or it can be reacted with sodium hydroxide to form sodium hypochlorite (bleach) which can then be added to purify water. The use of chlorine to purify drinking water is responsible for saving millions from death and sickness due to water-borne diseases such as typhoid, cholera, dysentery, and gastro-enteritis.

$$Cl_2 + 2\ NaOH \longrightarrow NaClO + NaCl + H_2O$$

Sodium hypochlorite

Sodium hydroxide has many applications in the chemicals industry and is also used in the manufacture of soaps, detergents, textiles, and aluminum. Often the demand of the caustic end-use markets varies from the demand of the chlorine end-use markets. However, because they are coproduced, no one makes caustic without making chlorine (or vice versa), even if they only have orders for caustic. Because of these market fluctuations, the chloralkali business necessitates a balance not required for other chemical manufacture. To produce either caustic or chlorine, there needs to be an outlet for the other.

Chlorine is a severe eye, nose, throat, and upper respiratory tract irritant and breathing high concentrations can be fatal [26]. At the Battle of Ypres, Germany used chlorine as an asphyxiant against the French in World War I, killing an estimated 800 [27]. Because of the problems inherent in shipping a toxic compressible gas, much of the chlorine produced is produced captively. Captive production means that a chemical is made for internal use and not for transportation and sale. By manufacturing chlorine and then consuming it at the same location, transportation risks can be avoided. Safety considerations also dictate against the storage of large quantities of chlorine. Many of the large chlorine producers use it directly to make chlorinated products such as EDC.

QUESTIONS

1. A container of 25% oleum could be considered ________% sulfuric acid (to nearest tenth of a percent).

2. Assuming 100% yield, a chloralkali plant that produces 320 tons of caustic soda, also produces ________ tons of ________ *(answer based upon the larger volume chemical that is coproduced)*

3. Which of the following is the largest use of sulfuric acid?

a. mining

b. waste water treatment

   c. fertilizers
   d. plastics manufacture

4. A plant that consumes 585 tons of salt produces how many tons of caustic (sodium hydroxide)?
   a. 298
   b. 400
   c. 585
   d. 710

5. What are the two gases that are coproduced in a caustic plant?

6. Why were there so many chloralkali plants built in Niagara Falls, NY?

REFERENCES

1. Lori E Apodaca. Sulfur 2009, U.S. Geological Survey Minerals Yearbook 2009.
2. Daum, KH, et al. U.S. Pat. Appl. Pub. No. 2008/0145290. 2008.
3. Schoubye, PCS. U.S. Pat.No. 4,348,373. 1982.
4. Perkin Medal Acceptance by Herman Frasch. The Journal of Industrial and Engineering Chemistry 1912 Feb: 134–140.
5. ND Ganguly, AC Banerjeel Ind. Eng. Chem. Process Des. Develop. 1973; 12(1).
6. Dolan, WB, et al. U.S. Pat. No. 7,311,891. 2007.
7. Lori E Apodaca. U.S. Geological Survey, Mineral Commodity Summaries, January 2012.
8. SM Jasinski. U.S. Geological Survey, Mineral Commodities Summaries, Jan. 2012.
9. SM Jasinski. Phosphate Rock U.S. Geological Survey, 2010 Minerals Yearbook.
10. Ishikawa, K, et al. U.S. Pat. No.7,470,414. 2008.
11. Meng, X, et al. U.S. Pat. No. 7,374,740. 2008.
12. RA Hutchins. Advances in Phosphate Fertilizer Technology 1993; 292: 70–80.
13. RA Hutchins. Advances in Phosphate Fertilizer Technology 1993; 292: 70–80.
14. MM Miller. U.S. Geological Survey, Mineral Commodity Summaries, January 2012.
15. National Lime Association Fact Sheet: Properties of Lime, January 2007.
16. Dennis S. Kostick. U.S. Geological Survey, Mineral Commodity Summaries, January 2012.
17. O Ozdemir, A Jain, V Gupta, X Wang, JD Miller. Minerals Engineering 2010; 23:1–9.
18. Joseph Gambogi. U.S. Geological Survey, Mineral Commodity Summaries, January 2012.
19. Julia Burdge. *Chemistry*, Second Edition. McGraw Hill; 2011.

20. JH Braun, A Baidins, and RE Marganski.Progress in Organic Coatings 1992; 20: 105–138.
21. M Alatalo, K Heikkila. Titanium Dioxide. In: E Lehtinen, editor. *Pigment Coating and Surface Sizing of Paper*. Atlanta: Tappi Press; 2000.
22. M Akhtar, et al. U.S. Pat. No. 6,562,314. 2003.
23. Dennis S Kostick. U.S. Geological Survey, Mineral Commodity Summaries, January 2012.
24. Kenneth S Johnson. Salt Resources and Production in the United States, Forum on Geology of Industrial Minerals, 36th Bath, England; 2000.
25. Michael McCoy. Chemical and Engineering News 2010; 88(51):22.
26. William E Luttrell. Chemical Health and Safety January/February 2002: 24–25.
27. Richard B Evans. Lung 2004; 183:151–167.

CHAPTER 3

Gases

3.1 SYN GAS

Carbon monoxide and hydrogen are perhaps two of the simplest building blocks in chemistry. Carbon monoxide can be converted to other one-carbon molecules such as methanol, used to make hydrocarbons as in the Fischer-Tropsch process, or used to extend molecules by addition of one carbon as in hydroformylation of alkenes. Hydrogen has many uses. It is used to make inorganic compounds such as ammonia or make hydrocarbons. Hydrogen has widespread use in the reduction of unsaturated organic compounds. Many processes rely on both carbon monoxide and hydrogen. Examples include the production of methanol, the Fischer–Tropsch process and hydroformylation reactions. In addition to often being used together, they are typically coproduced.

Syn gas, or synthesis gas [1, 2], is a term for mixtures of hydrogen and carbon monoxide. Coal was the major feedstock, but today the mixture is mainly made by the partial oxidation of hydrocarbons. The source of the hydrocarbon varies, but is commonly natural gas or petroleum distillates. The composition of natural gas varies depending on its source but it is mainly methane and can have other components such as hydrogen, ethane, propane, butane, and nitrogen. Often the petroleum distillates used for the production of syn gas are the heavier fractional residues left after the oil distillation in the gasoline refining process. The partial oxidation is exothermic and in the case of petroleum, gives a 1:1 molar yield of carbon monoxide and hydrogen.

$$\text{—CH}_2\text{—} + 0.5\ O_2 \longrightarrow CO + H_2$$

The exotherm can be controlled by employing a second endothermic reaction with steam.

Fundamentals of Industrial Chemistry: Pharmaceuticals, Polymers, and Business, First Edition. John A. Tyrell.

$$—CH_2— + H_2O \longrightarrow CO + 2\,H_2$$

The two reactions are run in conjunction and adjusted by varying the amount of oxygen and water. The reactions are run in such a manner as to satisfy the need to balance the heat requirements and also the desired H_2:CO ratio.

The syn gas process is usually run for specific purposes. Examples include feedstock for ammonia production, feedstock for methanol production, feedstock for the Fischer-Tropsch synthesis, or for the energy output. Depending on the reason for running the reaction, a different ratio of H_2:CO is needed. This ratio is influenced strongly by the material undergoing partial oxidation. Coal gives a 1:1 ratio of H_2:CO; petroleum about a 2:1 ratio and natural gas somewhat higher than a 2:1 ratio as will be evident from the discussion on methane oxidation.

Some of the principles can be illustrated with the partial oxidation of methane. Methane is a major component of natural gas so this reaction is important when natural gas is the feedstock.

$$2\,CH_4 + O_2 \longrightarrow 2\,CO + 4\,H_2$$

This is an exothermic reaction. For processes occurring at constant pressure, the heat flow can be expressed as an enthalpy change. Enthalpy (H) is the internal energy of the system plus the product of pressure and volume.

$$H = E + PV$$

At constant pressure, the change in enthalpy equals the heat gained or lost. We can calculate standard enthalpy changes of a reaction from standard enthalpies of formation. Standard state is 1 bar pressure and a temperature of 298 °K (25 °C) and is symbolized by the degree sign after H as H°.

$$\Delta H^{\circ}_{rxn} = \sum n\Delta H^{\circ}_{f}(\text{products}) - \sum n\Delta H^{\circ}_{f}(\text{reactants})$$

The standard enthalpy of formation of a compound is the enthalpy change for the reaction that forms one mole of the compound from its elements with all substances in their standard states. Enthalpies of formation are readily available. For $CO_{(g)}$, $\Delta H^{\circ}_{f} = -110.5$ kJ/mol. For $CH_{4(g)}$ $\Delta H^{\circ}_{f} = -74.8$ kJ/mol. Oxygen gas and hydrogen gas are each elements in their standard state so $\Delta H^{\circ}_{f} = 0$. For the partial oxidation of 2 moles of methane gas at standard temperature and pressure, we can calculate the enthalpy change to be:

$$2(-110.5\ \text{kJ}) - 2(-74.8\ \text{kJ}) = -71.4\ \text{kJ}$$

$$2\ CH_{4(g)} + O_{2(g)} \longrightarrow 2\ CO_{(g)} + 4\ H_{2(g)} \quad \Delta H^\circ = -71.4\ \text{kJ}$$

A second reaction for the oxidation of methane is steam reforming.

$$CH_{4(g)} + H_2O_{(g)} \longrightarrow CO_{(g)} + 3\ H_{2(g)}$$

For $H_2O_{(g)}$, $\Delta H^\circ_f = -241.8$ kJ/mol. For this reaction, we can calculate the standard enthalpy change to be

$$((-110.5\ \text{kJ})) - ((-74.8\ \text{kJ}) + (-241.8\ \text{kJ})) = 206.1\ \text{kJ}$$

$$CH_{4(g)} + H_2O_{(g)} \longrightarrow CO_{(g)} + 3\ H_{2(g)} \quad \Delta H^\circ = 206.1\ \text{kJ}$$

This reaction is highly endothermic. By varying the amount of oxygen and water, the two reactions can be done together in a manner so that the heat flow is controlled. The enthalpy calculations are not as simple as shown here because the reactions are not done at standard conditions and enthalpy changes with pressure and temperature. High pressures and temperatures of about 1200 °C are common. For precision, it needs to be calculated or experimentally determined for the conditions of interest.

For steam reforming, the carbon monoxide can in turn be oxidized by reaction with water to form carbon dioxide and hydrogen.

$$CO_{(g)} + H_2O_{(g)} \longrightarrow CO_{2(g)} + H_{2(g)}$$

For $CO_{2(g)}$, $\Delta H^\circ_f = -393.5$ kJ/mol. For this reaction, we can calculate the standard enthalpy change to be

$$(-393.5\ \text{kJ}) - (-110.5 + (-241.8\ \text{kJ})) = -41.2\ \text{kJ}$$

$$CO_{(g)} + H_2O_{(g)} \longrightarrow CO_{2(g)} + H_{2(g)} \quad \Delta H^\circ = -41.2\ \text{kJ}$$

When the two steam reforming reactions are combined, the overall reaction converts one mole of methane to 4 moles of hydrogen.

$$CH_{4(g)} + 2\ H_2O_{(g)} \longrightarrow CO_{2(g)} + 4\ H_{2(g)}$$

Most hydrogen made in the United States is made by this steam reforming process. One mole of methane can be converted to 4 moles of hydrogen. Taking into account molecular weights, 16 g of methane is converted to 8 g of hydrogen. This mass ratio of 2:1 works for other mass units. Ten tons of methane can yield 5 tons of hydrogen. Syn gas is used as a chemical feedstock.

The hydrogen is used for many processes, but ammonia synthesis is a major use. Another major use of hydrogen is in oil refineries to upgrade different petroleum fractions. Hydrogen is also used in the reduction of many organic chemicals. Carbon monoxide is used for several processes, but two major uses are with methanol to prepare acetic acid and with chlorine to prepare phosgene.

$$CO + CH_3OH \longrightarrow \underset{\text{Acetic acid}}{CH_3CO_2H} \qquad CO + Cl_2 \longrightarrow \underset{\text{Phosgene}}{COCl_2}$$

Syn gas is also used to produce methanol, to produce hydrocarbons by the Fischer–Tropsch synthesis and to react with olefins in the synthesis of aldehydes and alcohols by hydroformylation. For the synthesis of methanol, syn gas is reacted, typically with a copper-based catalyst to form methanol by the following equation.

$$CO + 2\,H_2 \rightleftharpoons CH_3OH$$

The reaction shown is an oversimplification and items such as shifting the equilibrium toward methanol, catalyst poisoning, and pressure/temperature tradeoffs need to be considered and are the subject of a text [3]. Methanol has many uses as a solvent and is the major component of windshield washer fluid. However, most methanol is converted to other commodity chemicals such as acetic acid, formaldehyde, or dimethyl terephthalate. It can also be used directly as a fuel additive, as a component in the synthesis of biodiesel or converted to dimethyl ether or methyl-t-butyl ether (MTBE), two fuel additives.

CH_3OCH_3
Dimethyl ether
CO
CH_3CO_2H
Acetic acid
CH_3OH
[O]
CH_2O
Formaldehyde
$(CH_3)_3C{-}O{-}CH_3$
MTBE
CO_2CH_3
CO_2CH_3
Dimethyl terephthalate

Celanese has announced [4] that they have developed a new technology enabling them to produce ethanol from synthesis gas. They project a significant cost improvement versus corn-based ethanol and are planning two plants in China, each with 400,000 metric ton capacity and at a cost of $300 million. They have also announced the construction of a technology development plant in Clear Lake, Texas.

In the Fischer–Tropsch synthesis, carbon monoxide reacts with hydrogen and is reduced to a hydrocarbon.

$$n\,CO + (2n+1)H_2 \longrightarrow H\left(CH_2\right)_n H + n\,H_2O$$

The reaction is done with a metal catalyst. Catalysts based upon iron or cobalt have been used commercially for hydrocarbon synthesis [5]. The mechanism involves adsorption of hydrogen and carbon monoxide on the metal surface [6]. The Fischer–Tropsch process enables natural gas to be converted to liquid synthetic fuel. First, the natural gas is oxidized to syn gas which is then converted by the Fischer–Tropsch process to the liquid hydrocarbon mixture, which is useful as fuel.

Hydroformylation involves the reaction of an alpha-olefin with carbon monoxide and hydrogen and a catalyst to form the aldehyde. Common catalysts are based upon cobalt or rhodium. The reaction is also referred to as the Oxo process or the Roelen reaction. The reaction is illustrated below with propylene as the alkene. The linear: branched isomer formation depends on catalyst selection with most companies trying to maximize the n-butyraldehyde because it has more uses. However the desired production ratios vary with market demand for end-use products.

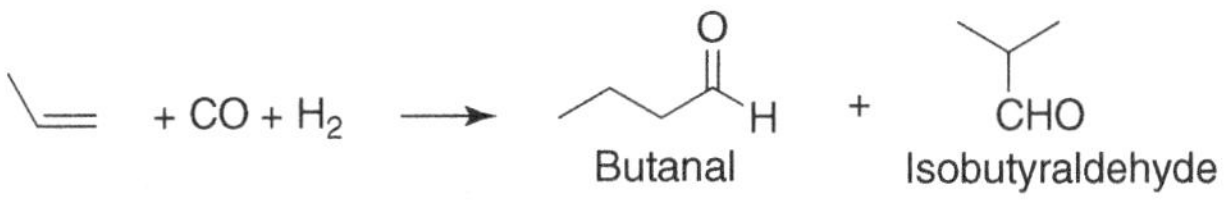

+ CO + H2 → Butanal + Isobutyraldehyde (CHO)

The aldehydes can be the desired product but often they are reduced to the alcohol or oxidized to the carboxylic acid. Other reactions, such as aldol condensations, can be employed as illustrated with butanal. Reduction of the aldol product gives the commodity chemical, 2-ethylhexanol, used to make plasticizers such as dioctyl phthalate (dioctyl phthalate is the common name for the diester of phthalic acid with 2-ethylhexanol, but more precisely, it can be called di-2-ethylhexyl phthalate).

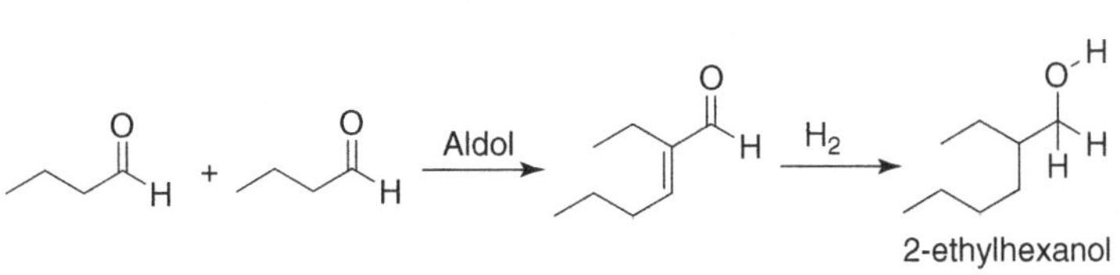

Aldol → H2 → 2-ethylhexanol

The components of syn gas can be separated by cryogenics. At low temperatures, any water or carbon dioxide that is present is a solid and can be easily removed. Carbon dioxide, which can be present because of over-oxidation, can also be removed by passing the gas through a base which traps the acidic carbon dioxide. This type of process is called scrubbing. Both carbon dioxide and water can also be removed by using molecular sieve adsorbents. There is about a 60 degree difference in the atmospheric boiling points of hydrogen (−253 °C) and carbon monoxide (−192 °C) and the two components can be separated by a low temperature pressure distillation.

Another technique to separate a mixture of hydrogen and carbon monoxide is to, under pressure, selectively adsorb the carbon monoxide on zeolites or some other adsorbent. Purified hydrogen passes through the adsorbent. Release of the pressure then desorbs the carbon monoxide. This is called "Pressure Swing Adsorbtion" or PSA.

Zeolites are aluminosilicates of group 1A and 2A elements such as sodium, magnesium, or calcium. They can be represented by the empirical formula:

$$M_{2/n}O \cdot Al_2O_3 \cdot ySiO_2 \cdot wH_2O$$

where n is the cation valence, y is typically 2–10, but can be higher for high silica zeolites, and w represents the number of water molecules in the voids [7]. They are crystalline and the AlO_4 or SiO_4 linkages form tetrahedra which in turn form rings that are linked together to give a porous crystalline structure. Depending on the type of zeolite, different numbers of rings are connected to form channels or pores of different sizes. For example, zeolite A has eight connected rings, resulting in a pore size of 0.35–0.45 nm. Zeolite Y has 12 connected rings and a pore size of 0.6–0.8 nm [8]. Because of the defined pore size, a specific zeolite can selectively fit molecules of a certain size. Due to this feature, zeolites are sometimes called "molecular sieves" and can be used to selectively adsorb molecules of one size or shape while others pass through unadsorbed.

Zeolites are naturally occurring, and were discovered in 1756 by the Swedish mineralogist Cronstedt, who noticed that they released water upon heating. Because of this property [9], he gave them the name "zeolite", which comes from the Greek for "boiling stones". Zeolites can also be synthesized and this allows chemists to tune properties such as charge and pore size. The International Zeolites Association has registered 179 different zeolite structures [10]. Zeolites are used as detergent builders to bind cations such as calcium and magnesium. They are used extensively as catalysts, especially in petroleum refining processes such as fluid catalytic cracking (FCC), reforming, and hydrotreating. They are used to separate mixtures. For example, the UOP process takes advantage of the size difference among p-xylene and its isomers to separate p-xylene. In the context of this discussion, zeolites can be used to separate carbon monoxide and hydrogen. Zeolites look like powder, but sometimes are extruded into cylindrical pellets.

A third separation technique relies on membranes. The syn-gas feed stream is passed over a semi-permeable membrane. A membrane is a layer of material that selectively allows one component to permeate through. Hydrogen selectively permeates the membrane and the retentate is enriched in carbon monoxide [11]. There are many materials used for membranes, including various polymers. Imagine a long plastic tube (think of a garden hose) that is made with a polymer that selectively allows hydrogen to permeate. We feed a pressurized mixture of hydrogen and carbon monoxide through the tube. Hydrogen passes through the tube walls as the gas travels through the tube. The gas exiting the end of the tube is enriched in carbon monoxide. A membrane module might consist of thousands of these tubes in the form of hollow fibers. The permeation rate through a membrane varies based on molecule size and solubility in the membrane material. Typically, smaller, more soluble materials permeate at a faster rate. Different plastics have been studied for this use with the optimum ones being those that allow one gas to selectively permeate through the plastic walls. The size of the molecule and the solubility of the gas in the plastic are factors in permeation rates with smaller soluble gases permeating faster. If you have ever stored a plastic soda bottle for a long period of time, you might have noticed that the soda loses its fizz. That is because the carbon dioxide has permeated through the plastic.

There are many types of materials that can be used for membranes useful for separating carbon monoxide from hydrogen. Material cost, chemical and thermal stability, and overall durability are important factors in the selection process. Hydrogen permeability and selectivity through the membrane are critical. Membrane materials have been reviewed [12] and can be categorized as metallic (pure metals or alloys), inorganics (including oxides, zeolites, glasses, and ceramics), porous carbons, purely organic polymers, and hybrids or composites.

Metallic membranes are often palladium alloys. They require the presence of specific catalytic surfaces to dissociate hydrogen on the raw feed stream side and reassociate the protons and electrons on the product side. Hydrogen selectivity is typically very high in these systems, but poisoning (rendering inactive) the metal surface with contaminants such as hydrogen sulfide is a major issue. Hydrogen can also cause metal embrittlement. Ceramics are also used as membranes. They are often based upon silica or coated silica and the separation principle is based upon molecular size as is the case with zeolites. Indeed, there are membranes based upon synthetic zeolites. Carbon-based membranes have been demonstrated but some forms suffer from high fabrication costs, brittleness, and degradation due to oxidation. Polymer-based membranes may the most economical to produce and easiest to fabricate, but often suffer from limited selectivity and low flux. There is continued research to modify the polymer structure to improve these properties.

For the separation of carbon monoxide and hydrogen, often a combination of techniques – cryogenics, PSA, and membranes – is used.

3.2 NITROGEN AND OXYGEN

Nitrogen and oxygen are prominent industrial gases and are obtained by separation from air. Air is about 78% by volume nitrogen, 21% oxygen, 0.9% argon and smaller amounts of other gases. Nitrogen is a colorless, odorless, tasteless gas with a specific gravity of 0.967 and a boiling point of −196 °C. The specific gravity of a gas is the density of the gas divided by the density of air. The density units cancel so specific gravity is unitless. The specific gravity of a liquid is the density of the liquid divided by the density of water.

Because nitrogen is ubiquitous, there is some familiarity with it, but it can be dangerous and can cause death if someone is in a nitrogen atmosphere with insufficient oxygen. I worked at one location when they held their 25th anniversary of operation. The site manufactured several toxic gases: carbon monoxide, phosgene, and chlorine, yet the two fatalities in the site's operating history were not from these, but from nitrogen. In a tragic accident, a worker went into an area that had been blanketed with nitrogen and a second worker was also overcome when he went to rescue the first worker. More recently, in 2011 at a Shintech vinyl chloride plant in Plaquemine, LA, two workers died after entering a vessel that had a nitrogen atmosphere [13]. Nitrogen is commonly used industrially to provide inert atmospheres and prevent fires, but precautions must be taken before entering these atmospheres. There is also the danger associated with any pressurized gas. In 2013, at a CF Industries plant in Louisiana that manufactured ammonia, a manifold ruptured during off-loading of nitrogen gas. One worker was killed and seven others were injured [14].

Oxygen has a boiling point of −183 °C which is sufficiently different from nitrogen that the two components can be separated from air by a cryogenic distillation. Nitrogen has the lower boiling point and exits from the top of the fractionating column and oxygen from the bottom. In laboratory usage of liquid nitrogen, such as in a Dewar flask, care should be taken to avoid condensation of oxygen which can accumulate if the Dewar is continually "topped off" with more liquid nitrogen rather than allowed to evaporate to dryness between fillings. This can eventually lead to oxygen continuing to condense and accumulate in the open Dewar with the end result being a flammable or explosive level of oxygen.

After cryogenic distillation of air, the fractionated nitrogen and oxygen have sufficient purity for most applications. The argon, with a boiling point −186 °C is in the oxygen fraction. The argon can be separated and the oxygen purified by more elaborate fractionation. Other noble gases such as neon, krypton and xenon can also be isolated by the distillation of air. The primary source of helium is from natural gas wells.

Oxygen and nitrogen can also be separated from air by PSA or by membranes. Depending on the requirements, a combination of these techniques is used. Both stationary and portable home oxygen concentrators based upon PSA are used for patients prescribed oxygen therapy [15, 16]. A combination of membrane technology with PSA can also be used [17].

For membrane separation of nitrogen and oxygen, typically the oxygen has a high permeation rate and the nitrogen a low permeation rate. Pressure Swing Adsorption separation uses zeolites. In the separation of air, typically the oxygen molecules are adsorbed under pressure in the channels of the zeolite and the nitrogen passes through. Upon release of the pressure, the oxygen is desorbed and more air can be introduced into the vessel containing the zeolites. By varying (pressure swing) the pressure, oxygen is selectively adsorbed and then desorbed, allowing it to be separated from air.

Oxygen is used in the production of steel and in the chemical industry as an oxidant. Examples where oxygen is used as an oxidant include the reaction with ethylene to form ethylene oxide, the conversion of methanol to formaldehyde, and the oxidation of cumene to form, ultimately, phenol and acetone. Often, such as in the reaction of cumene, the oxygen content of air is sufficient and the oxygen does not have to be concentrated.

$H_2C{=}CH_2 \xrightarrow{O_2}$ Ethylene oxide

$CH_3OH \xrightarrow{O_2} CH_2O$ Formaldehyde

Cumene $\xrightarrow{O_2}$ (O–O–H) $\xrightarrow{H^+}$ Phenol + Acetone

The largest use for nitrogen is to produce ammonia. However nitrogen also has many industrial uses such as to provide an inert atmosphere in chemical and metallurgical processes and, in liquid form, as a coolant. In nature, enzymes convert atmospheric nitrogen into ammonia in a process known as nitrogen fixation. Nitrogen fixation serves as a source of nitrogen for proteins and other naturally-occurring molecules.

3.3 AMMONIA

The industrial conversion of nitrogen to ammonia is done by a process known as the Haber process. In 1908, Fritz Haber invented a process to convert nitrogen and hydrogen in the presence of an iron catalyst into ammonia. In 1919, he was awarded the 1918 Nobel Prize in Chemistry – a year later due to suspension of Nobel Prizes during World War 1. One motivation of Haber's research was to provide Germany with a source of ammonia for the production of explosives [18]. Previously, the main source of nitrogen compounds had been from Chilean saltpeter, sodium nitrate, which was blockaded in World War I. Haber was also involved in the development of the chemistry of poisonous gas warfare. The year after he won the Nobel Prize, he was charged with international war crimes for his role in chemical warfare during World War I.

Later, Carl Bosch developed the process on an industrial scale. The Nobel Prize in Chemistry 1931 was awarded jointly to Carl Bosch and Friedrich Bergius for their contributions to the invention and development of chemical high pressure methods.

The process is called the Haber process or sometimes, the Haber–Bosch process. It is estimated that the number of humans supported per hectare of arable land has increased from 1.9 to 4.3 persons between 1908 and 2008 mainly due to fertilizers provided by the Haber–Bosch process and by 2008, nitrogen fertilizers were responsible for feeding 48% of the world's population [19]. Most ammonia produced is used for fertilizers, either by direct application or by conversion to solid fertilizers such as urea, ammonium nitrate, or ammonium phosphate. Nitrogen is a primary nutrient for plants and large amounts of nitrogen are needed for agriculture. Other uses of ammonia include the manufacture of chemicals such as acrylonitrile or caprolactam, and miscellaneous uses such as pulp and paper and refrigeration.

$$N_2 + 3H_2 \rightleftharpoons 2NH_3 + 92\ \text{kJ}$$

The process is an equilibrium. Ammonia formation is favored by cold temperatures and high pressures. However, low temperature means slower rate of ammonia production and added costs associated with cooling the exothermic reaction. Higher pressure also necessitates higher costs in reactor design and compression pumps. For example, higher pressures require reciprocating pumps, which are like bicycle floor pumps, versus centrifugal pumps, which can be used at lower pressures and operate on a fan-blade type principle. A variety of catalysts can be used to improve the rate and the introduction of copper catalysts has enabled somewhat lower process temperatures [20]. However, an ammonia process is a balance between conversion and cost. A typical process might run at temperatures from about 350 to 550 °C and pressures of 120 to 250 atmospheres. Under these conditions, the conversion rate may be 30% or less. You may wonder how it can be economical to produce ammonia with such a low conversion rate. It is because the product ammonia is readily separated from the unreacted hydrogen and nitrogen. This can be done by cooling. Ammonia has a boiling point of −33 °C but nitrogen and hydrogen have boiling points of −196 °C and −253 °C respectively. Therefore the ammonia can be selectively condensed and unreacted nitrogen and hydrogen gas returned to the reactor as a recycle stream. Because of the recycling of unreacted starting material, the overall reaction yield is very high, despite the low conversion (amount of product per pass).

The reaction is shown with an exotherm of 92 kJ. Expressed as the exotherm per mole of ammonia, it is 46 kJ. This is at standard temperature and pressure. Although this is fine for most academic discussions, one should realize that enthalpy changes are influenced by both temperature and pressure. Ammonia is not commercially made at standard temperature and pressure (a temperature of 273.15 °K, 0 °C and a pressure of 1 atmosphere; 1 atmosphere = 1.01 bar = 0.101 MPa) so heats need to be either calculated or experimentally determined at the conditions of the synthesis. For example, at atmospheric pressure with an increase in temperature from 300 °K to 600 °K to 900 °K, the enthalpy change per mole of ammonia is increased from 46.35 kJ to 52.04 kJ to 55.06 kJ [21]. The enthalpy change increases with increasing pressure. For example, at 400 °C, it is 53.09 kJ/mol NH_3 at 10 MPa. At the same temperature and 40 MPa, it increases to 57.18 kJ. At 100 MPa, it is 65.37 kJ [22].

For a plant operating at 400 °C and 10 MPa, somewhat typical commercial conditions, we can calculate the enthalpy change per metric ton of ammonia produced. A ton is 2,000 pounds and a metric ton, sometimes referred to as "tonne" is 1,000 kg which is 2,200 pounds.

$$(53.09\ \text{kJ/mol NH}_3)(1\ \text{mole/17g})(1000\ \text{g/kg})(1000\ \text{kg})$$
$$= 3.12\ \text{million kJ}\ (3.12 \times 10^9\ \text{J}).$$

The reaction uses a catalyst, typically an iron catalyst. Magnetite (Fe_3O_4) is used with other oxides such as alumina, calcium oxide, potassium oxide, magnesium oxide and silica. The magnetite is reduced in the process to the active iron catalyst [23]. Alumina acts as a structural stabilizer and gives improved surface area in the catalyst. Particle size must be balanced. Smaller particles are more effective catalysts but very small particles can cause pressure drops through the reactor. Certain substances may interfere with the catalyst. These are called poisons. In general, a catalyst that is resistant to poisons is desirable, but there must still be efforts to exclude poisons from the reactor feed.

Although Haber invented his process more than 100 years ago, it is still very much an important process. For example, in late 2012 several companies announced plans for capacity expansion or new facilities [24]. Illinois-based CF Industries announced it will spend $3.8 billion through 2016 on new ammonia and ammonia derivative capacity at its plants in Louisiana and Iowa. Orascom Construction Industries announced plans for a $1.4 billion nitrogen fertilizer plant in Iowa and CHS Inc. announced plans for a $1 billion plant in North Dakota.

One major derivative of ammonia is nitric acid. Nitric acid is used for fertilizer applications such as to make ammonium nitrate or potassium nitrate. It also has widespread use in organic chemistry especially in the nitration of aromatic rings. Nitric acid is made by the Ostwald process. Ammonia is oxidized first to nitrogen oxide and then to nitrogen dioxide, followed by absorption in water resulting in nitric acid. Nitric acid forms an azeotrope with water at 68% nitric acid. Because of this, it cannot be purified further by simple distillation. More concentrated grades of nitric acid can be prepared by use of a dehydrating agent such as sulfuric acid and distillation. Fuming nitric acid is prepared by passing nitrogen dioxide through concentrated nitric acid. Depending on the concentration of nitrogen dioxide, the solution takes on a yellow to reddish-brown color.

The oxidation of ammonia proceeds through a series of steps. In the first step, ammonia is oxidized with a platinum catalyst. The catalyst is usually used in the form of a gauze made from a platinum–rhodium alloy. Compared with platinum, an alloy of platinum with 5 to 10% rhodium improves the mechanical strength of the gauze, improves the yield, and reduces the activity loss of the catalyst [25]. Typically a knitted gauze with three-dimensional geometry and high porosity is used. As ammonia is oxidized, platinum is slowly lost from the gauze. After ammonia, catalyst consumption is the second largest raw material cost. Ammonia is oxidized commercially at pressures ranging from one to ten atmospheres and temperatures ranging from

815 °C to 930 °C with most plants operating at about 10 atmospheres [26]. The temperature is chosen by balancing nitric oxide yield, which increases with increasing temperature, and catalyst loss, which also increases with increasing temperature. Commonly, the oxidation takes place at about 900 °C.

$$4\,NH_{3(g)} + 5\,O_{2(g)} \xrightarrow{\text{Catalyst}} 4\,NO_{(g)} + 6\,H_2O_{(g)}$$

$$2\,NO + O_{2(g)} \longrightarrow 2\,NO_{2(g)} \rightleftharpoons N_2O_4$$

$$3\,NO_2 + H_2O \longrightarrow 2\,HNO_3 + NO$$

The nitric oxide which is formed reacts with oxygen to form nitrogen dioxide. Nitrogen dioxide exists in equilibrium with its dimer, dinitrogen tetroxide. The nitrogen dioxide/dimer mixture is sent to a column, sometimes called an absorption tower. Water is added at the top of the column. The nitrogen dioxide is converted to nitric acid. Byproduct nitric oxide is oxidized to nitrogen dioxide by means of a stream of air passed into the absorption column. The aqueous nitric acid is removed continuously from the base of the column. Overall, the reaction can be written as:

$$NH_3 + 2\,O_2 \longrightarrow HNO_3 + H_2O$$

Industrial processes are single-pressure or two-pressure. In the single-pressure process, the absorption takes place at the same pressure as the catalytic oxidation. In the two-pressure process, the absorption pressure is higher than the catalytic oxidation. The catalytic oxidation (first reaction) is favored by low pressures and the two subsequent reactions (nitric oxide oxidation and reaction with water) by high pressure. Therefore, newer plants tend to operate with the two-pressure process. The main advantage is less platinum consumption.

Recognize that the reaction to convert ammonia to nitric acid is more complicated than summarized here. One set of authors proclaimed, "Never has a reaction system been studied so long with ultimately so little consensus on mechanistic explanation [27]." For example, ammonia can be oxidized to form nitrogen or nitrous oxide (N_2O).

$$4\,NH_3 + 3\,O_2 \longrightarrow 2\,N_2 + 6\,H_2O$$

$$2\,NH_3 + 2\,O_2 \longrightarrow N_2O + 3\,H_2O$$

At relatively low temperatures (200 °C), nitrogen is the only product. At higher temperatures, nitrous oxide formation begins passing through a maximum at 400 °C [28]. The desired nitric oxide (NO) begins to be formed at 300 °C and the yield of nitric oxide continues to increase with temperature. Most plants operate at about 900 °C. However, even at this temperature, both nitrogen and nitrous oxide are byproducts.

At the stoichiometric ratio of ammonia to oxygen, the ammonia/air mixture is explosive. Therefore lower levels of ammonia are used. The lower explosion level (LEL) declines with pressure so low pressure systems can use a higher level of ammonia than high pressure systems [29]. There are other potential explosive hazards. For example, formation of ammonium nitrite or ammonium nitrate is a concern and precautions must be taken in order to avoid explosions.

Nitric acid is used to make nitrate salts for fertilizer application. Ammonium nitrate and potassium nitrate are common. Ammonium nitrate is also used for explosives. Nitric acid is used in the production of adipic acid which in turn is used to make nylon 6,6 and also some polyesters. In the synthesis of adipic acid, cyclohexane is oxidized by a radical process to a mixture of cyclohexanol and cyclohexanone. This mixture is then oxidized further with nitric acid and a catalyst to adipic acid.

O_2, Catalyst → Cyclohexanol + Cyclohexanone

Cyclohexanol + Cyclohexanone → (HNO_3, Catalyst) Adipic acid

Nitric acid is also used in the nitration of aromatic rings. For example, if toluene is nitrated twice to form dinitrotoluene, the intermediate mixture of predominantly 2,4-dinitrotoluene and 2,6-dinitrotoluene can be reduced to the corresponding diamines which are useful for polyurethane manufacture. If three nitro groups are added, the explosive trinitrotoluene (TNT) is formed. Recognize that because each nitro group introduced withdraws electron density from the aromatic ring, the reaction rate slows by several orders of magnitude for each nitro group introduced. Because of this, it is possible to selectively add one, two, or three nitro groups to synthesize different target molecules. The introduction of the second nitro group requires higher temperatures and the third, higher still.

CH_3 $\xrightarrow[H_2SO_4]{HNO_3}$ O_2N, CH_3, NO_2 + CH_3, NO_2, NO_2

O_2N, CH_3, NO_2 + CH_3, NO_2, NO_2 $\xrightarrow{\text{Reduction}}$ H_2N, CH_3, NH_2 + CH_3, NH_2, NH_2

CH_3 $\xrightarrow[H_2SO_4]{HNO_3}$ O_2N, CH_3, NO_2, NO_2

TNT

QUESTIONS

1. What is the specific gravity of nitrogen?
   - **a.** 0.967 g/mL
   - **b.** 0.967 g/cm^3
   - **c.** 0.967 g/gallon
   - **d.** 0.967 g/ft^3
   - **e.** 0.967
2. Name two ways to produce nitrogen.
3. What word derives from the Greek for "boiling stones" and refers to crystalline aluminosilicates having well-defined pore structures?
4. Which statement is true (choose the best answer).
   - **a.** Reaction conversion is often higher than reaction yield
   - **b.** Reaction conversion is always the same as reaction yield
   - **c.** Conversion is the amount of product formed per pass through the reactor
   - **d.** Conversion is the amount of recycle returned to the reactor.
5. Assuming 100% yield, an ammonia plant that consumes 140 tons of nitrogen produces how much ammonia?

a. 140 tons
b. 170 tons
c. 280 tons
d. 340 tons

6. An ammonia plant typically operates with a ________ of less than 40%, but a ________ of greater than 90%. The two missing words (in order) are?
   a. purity ... conversion
   b. yield ... conversion
   c. conversion ... yield
   d. yield ... purity

7. Synthesis gas is a term used for:
   a. nitrogen and oxygen
   b. carbon monoxide and hydrogen
   c. benzene/toluene/ xylene or BTX
   d. ammonia, nitrogen, and hydrogen

8. Between the two, which is the higher boiling component of air (less volatile)?
   a. nitrogen
   b. oxygen

9. If a plant operating at 400 °C and 40 MPa produces 5 metric tons of ammonia, what is the amount of heat generated?

10. By utilizing steam reforming, what is the maximum amount of hydrogen that can be produced from a process that consumes 300 tons of methane?

11. Calculate the standard enthalpy change for:

$$CH_{4(g)} + 2H_2O_{(g)} \longrightarrow CO_{2(g)} + 4H_{2(g)}$$

12. Show the reaction sequence to prepare n-butanol using syn gas.

REFERENCES

1. K Weissermel, H-J Arpe. *Industrial Organic Chemistry*. 3rd ed. New York: VCH Publishers Inc.; 1997:13–23.
2. H Wittcoff, B Reuben. *Industrial Organic Chemicals*. New York: Wiley-Interscience; 1996:308–311.
3. S Lee. *Methanol Synthesis Technology*. Boca Raton, FL: CRC Press; 1990.

4. Alexander Tullo. Chemical and Engineering News 2011; 89(43):20.
5. AY Khodakov,W Chu, P Fongarland. Chem. Rev. 2007; 107:1692–1744.
6. CK Rofer-DePoorter. Chem. Rev. 1981; 81:447–474.
7. H Van Bekkum. *Introduction to Zeolite Science and Practice, Technology & Engineering*. Elsevier, Amsterdam, Netherlands; 2001:12.
8. H Van Bekkum. *Introduction to Zeolite Science and Practice, Technology & Engineering*. Elsevier, Amsterdam, Netherlands; 2001:13.
9. H Van Bekkum. *Introduction to Zeolite Science and Practice, Technology & Engineering*. Elsevier, Amsterdam, Netherlands; 2001:5.
10. W Vermeiren, J-P Gilson. Top Catal.2009; 52:1131–1161.
11. E Scharpf. U.S. Pat. No 6,568,206. 2003.
12. Nathan W Ockwig, Tina M Nenoff. Chemical Reviews 2007; 107(10): 4078–4110.
13. Alexander Tullo. Chemical and Engineering News 2011; 89(22):26.
14. C Bettenhausen, N Heidari, Chemical and Engineering News 2013; 91(25):5.
15. MP Czabala. U.S. Pat No. 6,478,857. 2002.
16. RD Whitley, GP Wagner, MJ Labuda. U.S. Pat.No. 7,350,521. 2008.
17. CL Staiger, MR Vaughn, AK Miller, CJ Cornelius. U.S. Pat. Appl. No. 20100116132. 2010.
18. JW Erisman, MA Sutton, J Galloway, Z Klimont, W Winiwarter. Nature Geoscience 2008; 1.
19. JW Erisman, MA Sutton, J Galloway, Z Klimont, W Winiwarter. Nature Geoscience 2008; 1.
20. DK Mukherjee, AK Bhaduri. Fertiliser News 1993; 38(12):23–31.
21. M Appl. *Ammonia, Principles and Industrial Practice*. New York: Wiley; 1999:19.
22. M Appl. *Ammonia, Principles and Industrial Practice*. New York: Wiley; 1999:22.
23. M Appl. *Ammonia, Principles and Industrial Practice*. New York: Wiley; 1999:39.
24. Melody Bomgardner. Chemical and Engineering News 2012; 90(46): 6.
25. J Dubois, P Smith. U. S. Pat. No. 8,263,036. 2012.
26. D Echegary, A Velloso, M Wagner. U.S. Pat.No. 6,165,435. 2000.
27. J Perez-Ramirez, F Kapteijn, K Schoffel, J Moulijn. Applied Catalysis B: Environmental 2003; 44:117–151.
28. J Perez-Ramirez, F Kapteijn, K Schoffel, J Moulijn. Applied Catalysis B: Environmental 2003; 44:123.
29. L Marzo. Nitric Acid Production and Abatement Technology including Azeotropic Acid. Proceedings No. 540, International Fertiliser Society, York, UK. 2004:1–27.

CHAPTER 4

Patents

A patent is a contract with the government where, in exchange for disclosing your invention, you are given an exclusive right to the invention for a period of time. The inventor has the right to exclude others from making, using, offering for sale, selling, or importing the invention. Patents do not give the patent owner a right to do or sell anything, but the ability to prevent others from making, using, or selling a product or process that falls within the claimed invention [1]. Unlike copyrights, trademarks, and trade secrets, patent rights do not exist without an issued patent. The rules vary somewhat for different countries. This discussion will cover the United States, but many of the principles are true for other parts of the world. The U.S. Patent and Trademark Office (USPTO) administers the patent law and issues patents on behalf of the government. The Patent Office dates back to 1802. Patent law is established in U.S. Code Title 35. This established the USPTO. Patent rules are published in Title 37, Code of Federal Regulations.

Patent rights are generally granted for twenty years from the date of filing the patent application with the USPTO, an agency of the U.S. Department of Commerce. The patent right excludes others from "making, using, offering for sale, or selling" the invention in the United States or "importing" the invention into the United States.

One notable exception to the 20-year term arises from the Hatch–Waxman Act, also known as the "Drug Price Competition and Patent Term Restoration Act of 1984" [2]. The statute enables the owners of patents on certain human drugs, food or color additives, medical devices, animal drugs, and veterinary biological products to add to the 20-year term some of the time lost while awaiting premarket government approval from a regulatory agency. This also provides that it shall not be an act of infringement, for example, to make and test a patented drug solely for the purpose of developing and submitting information for an Abbreviated New Drug Application (ANDA) [3]. This second provision enables generic drug companies to obtain bioequivalency

Fundamentals of Industrial Chemistry: Pharmaceuticals, Polymers, and Business, First Edition. John A. Tyrell.

data and other necessary information to file an ANDA. Because of this, generic versions of a drug come on the market shortly after patent expiration.

One other aspect of the Hatch–Waxman Act is that it entices generic manufacturers to challenge the validity of innovator-company patents. It grants the first generic manufacturer that successfully challenges a patent exclusivity for 180 days over other generic manufacturers. Patent litigation can be lengthy and expensive; often these patent disputes are settled out of court between the innovator company and the generic company. Sometimes the settlements involve the agreement by the generic company to delay manufacturing. When this happens, it can lead to challenges of antitrust violation.

There are three types of patents: utility, design, and plant. Utility patents are granted for a new and useful process, machine, article of manufacture, or composition of matter, or any new and useful improvement. A design patent covers an ornamental design for an article of manufacture. A plant patent may be granted to anyone who invents or discovers and asexually reproduces any distinct and new variety of plant. Chemical patents are utility patents. They may be for a new chemical, a new way of making an existing chemical, or an improvement to a known method of making a chemical.

There are some things which cannot be patented. Those include physical phenomena, abstract ideas, as well as literary and other artistic works. Further, in order to be patentable, an invention must satisfy three criteria. It must be new, it must be useful, and it must be non-obvious to one skilled in the art. If it does not satisfy each of these criteria, a person is not entitled to a patent. The criteria to be considered new or novel are listed in 35 U.S.C. 102. A person loses his right to a patent if the claimed invention was known or used by others or patented or described in a printed publication before the invention thereof by the applicant for patent, or the invention was patented or described in a printed publication or in public use or on sale more than one year prior to the date of the application for patent in the United States.

Historically, in the U.S. the first to invent has been awarded the patent. In areas of dispute, it has come down to a witnessed and dated laboratory notebook page. It is common for different companies, especially competitors, to work in the same area and to have the same invention. There has been more than one instance where rights to an important invention were awarded based upon laboratory notebooks. Because of this, prudent companies are strict with their employees about record-keeping. The Leahy–Smith America Invents Act (HR1249) was signed into law on September 16, 2011. One provision of this act is that effective March 16, 2013, the awarding of a patent is no longer to the first to invent but the first inventor to file. Awarding the patent to the first to file is consistent with the practice of most of the rest of the world.

A person loses their right to a patent if the invention was patented or described in a printed publication in this or a foreign country or in public use or sale in this country more than one year prior to filing the patent application. This is true even if it is the inventor who discloses the invention. For example, if Joe invents a new chemical and publishes a paper in J.A.C.S. and then 13 months later decides to file a patent application on the new chemical, he is not entitled to a patent. Similarly, consider the situation where Joe has a formulation for a new mosquito repellent and sells it for more than a year, to make sure that the market is profitable enough to warrant legal and filing fees. He has lost his right to a patent.

The second criterion, usefulness, is typically not problematic. However, useful is not a synonym for widely adopted or commercially viable. So an invention may not be commercially attractive but could still be useful.

The third criterion, that the invention must be non-obvious to one skilled in the art, can be more difficult. The term "skilled in the art" means someone having ordinary skill in the same area. If the invention is a new chemical synthesis, it means a synthetic chemist. As an extreme example of obviousness, consider the following hypothetical situation. It is well known that phenol substituted with any of methyl, ethyl, butyl, pentyl, and hexyl groups is useful to prevent hair-loss. No one has ever tested or described propyl substituted phenol. Joe decides to try propyl phenol and it works similarly to the other previously described alkyl phenols. Joe's neighbor is a physicist and tells Joe that he thinks this is surprising. Joe is not entitled to a patent because a skilled chemist would recognize propyl as an obvious variant. Rarely is the obviousness criterion so straightforward and it is important to consider the invention in light of what was known prior to the invention. Many non-obvious things seem obvious after the fact.

It is the inventor who applies for a patent. When more than one person contributes to the idea conception, they are each inventors or collectively they are co-inventors. When working for a company, often the company's legal department files the application on behalf of the inventor. Typically, the inventor assigns the patent rights to their company and the company pays filing and other fees, such as maintenance. The assignee is the owner of the patent rights and is a distinct entity from the inventor. On occasion, there is confusion about inventorship. An inventor invents or discovers a process, composition, machine, manufacture, or improvement thereof falling within the scope of one or more claims of a patent application. Idea conception and contribution to the idea conception are essential. The manager of the person who conceives of the idea and the person who carries out experimentation to demonstrate the idea are not co-inventors unless they contribute to the idea conception.

A patent application is filed with the USPTO. The inventor files the application. The inventor can do this independently but it is more common and more advisable to do so with the aid of a registered patent agent or patent attorney. The USPTO recommends that the inventor be represented by a registered agent or attorney. The USPTO maintains a register of people who have the legal, scientific, and technical qualifications necessary to render applicants for patents a valuable service. Certain of these qualifications must be demonstrated by the passing of an examination, often called the patent bar exam. Those admitted to the examination must have a college degree in engineering or physical science or the equivalent of such a degree.

The USPTO registers both attorneys at law and persons who are not attorneys at law. The former persons are referred to as "patent attorneys" and the latter persons as "patent agents." Both patent attorneys and patent agents are permitted to prepare an application for a patent and conduct the prosecution in the USPTO. Patent agents, however, cannot conduct patent litigation in the courts or perform various services which the local jurisdiction considers as practicing law. For example, a patent agent could not draw up a contract relating to a patent, such as a licensing agreement if the state in which he/she resides considers drafting contracts as practicing law.

It is important for the inventor to involve the assistance of a skilled attorney or agent because otherwise there is a high risk of making mistakes in the application which could result in the patent not being allowed or severely limited in protection. Often wording, and in particular wording of the claims, has important meaning which may be overlooked by an inventor. Imagine for example, an invention of a salad. The application may claim a salad consisting of lettuce, pears, gorgonzola cheese, and tomatoes. Or the application may claim a salad consisting essentially of lettuce, pears, gorgonzola cheese, and tomatoes. Or the application may claim a salad comprising essentially of lettuce, pears, gorgonzola cheese, and tomatoes. Ignoring novelty and obviousness arguments, for the sake of this example, let's assume the patent is allowed and issues. Can a competitor sell a salad made from lettuce, pears, gorgonzola cheese, and tomatoes and also croutons?

It depends upon the wording of the claim. The answer is yes, a competitor can sell the salad if "consisting of" was used in the claim. The answer is no if "comprising" was used in the claim. The term "comprising" is inclusive or open-ended and does not exclude additional, unrecited elements or method steps. The phrase "consisting of" excludes any element, step, or ingredient not specified in the claim.The phrase "consisting essentially of" limits the scope of a claim to the specified materials or steps "and those that do not materially affect the basic and novel characteristic(s)" of the claimed invention.

A patent application typically contains a background statement which gives the field of the invention and explains the state of the art prior to the invention and the need for the invention. It then gives a summary of the invention followed by a detailed description. There is typically a section that gives examples which illustrate the invention. In this section, sometimes there are comparative examples of items similar to the invention that can be used to illustrate the advantage of the inventive examples. The most important part of the application is the section of claims where the inventor states what he is claiming as his invention. Often there is a broad claim and then more limited claims based upon the broad claim. These more limited claims flowing from a broad claim are called dependent claims because they depend on an earlier claim.

As an example, consider the first four claims of a recently issued patent [4].

1. A method for processing UHMWPE for use in medical applications; the method comprising the steps of:

   combining UHMWPE with an antioxidant to form a blend having 0.01 to 3.0 weight percent of the antioxidant;

   processing the blend to consolidate the blend, the consolidated blend having a melting point;

   preheating the consolidated blend to a preheat temperature below the melting point of the consolidated blend; and

   irradiating the preheated consolidated blend with a total irradiation dose of between 25 kGy and 1000 kGy while maintaining the consolidated blend at a temperature below the melting point of the consolidated blend to form an antioxidant stabilized crosslinked UHMWPE blend.
2. The method of claim 1, wherein said combining step further comprises combining UHMWPE with an antioxidant to form a substantially homogenous UHMWPE blend having 0.01 to 3.0 weight percent of the antioxidant.
3. The method of claim 1, wherein said combining step further comprises mixing the UHMWPE powder with tocopherol by at least one of solvent blending, high shear mixing, precision coating, utilizing a fluidized bed, atomization, emulsion polymerization, electrostatic precipitation, wetting or coating of particles, and master batch blending.
4. The method of claim 1, wherein said combining step further comprises mixing the UHMWPE powder with tocopherol by high shear mixing.

Claim 1 is the broad or independent claim. It claims the invention in its broadest sense. Claims 2–4 are dependent claims depending on claim 1 and are narrower aspects of the invention of claim 1. Claim 1 requires a consolidated blend but claim 2 requires that it be homogeneous. Claim 3 restricts the antioxidant of claim 1 to a specific antioxidant, tocopherol. Claim 4 restricts the claimed invention to both tocopherol and a particular method of mixing.

How broadly to claim an invention in a patent application is an important decision. If you claim too narrowly you may open up the area for competitors to practice variants outside the claims. If you claim too broadly, there is a greater chance that there will be prior art that causes your claims to be rejected. Even if your claims are allowed, there is a possibility that your patent may be later invalidated.

When a patent application is filed with the USPTO, then it is considered patent pending. A patent examiner looks at the claims and can allow them. More commonly, the patent examiner either raises some objections, perhaps to wording, or disallows some or all of the claims. For example, an examiner may object to claims as being indefinite. If a claim states about 10% to about 20%, the examiner may object that the term "about" is indefinite. The applicant can either argue that the meaning is clear, perhaps by pointing to the wording in the detailed description. Alternatively, the applicant can amend the wording and remove the modifier "about."

If the examiner deems that there is more than one independent invention and that there is a burden on the examiner if they are not separated, the examiner may restrict his examination to a single set of claims and require the applicant to file a second application for the other claims that cover a second invention.

The examiner can reject the claims based upon novelty, usefulness, or obviousness. In doing so, the examiner typically recites the portion of patent law that forms the basis for the rejection. For example, a novelty rejection is a 102 rejection after 35 U.S.C. § 102, Conditions for patentability; novelty and loss of right to patent. An obviousness rejection is known as a 103 rejection. Such a rejection should consider: the scope and content of the prior art, the differences between the prior art and the claimed invention, the level of ordinary skill in the pertinent art at the time the invention was made, and whether any objective evidence of non-obviousness exists. The examiner can combine several references when making an obviousness argument but it should be reasonable for one skilled in the art to combine the references.

If a patent application is rejected by the examiner, the applicant has an opportunity to respond. For example, in an obviousness rejection the applicant can argue that it is not reasonable to combine references or that the references do not show what the examiner has concluded they show. The applicant needs to show that the invention is non-obvious in light of the prior art. Arguments against a 103 rejection might include unexpected results, commercial success, long-felt need, or the failure of others. Sometimes claims are amended or cancelled based upon the examiner's rejection. The examiner considers the applicant's arguments and then either allows the claims, issues a second rejection, or issues a new rejection. This process involving the back and forth arguments between the examiner and the applicant is called the patent prosecution. The time from the initial filing of the application to the issuance of an allowed patent can be several years.

Patent protection is a critical element for the success of many corporations from large multi-national companies to small operations. With patent protection, companies can get a return from the investment they have made in research and can maintain a competitive advantage. The large number of patents applied for and issued each year is a testimony to the value of patent protection. In 2011, the USPTO issued 224,505 utility patents. That same year, just over 500,000 patent applications were submitted. International Business Machines Corp. led all companies with 6,148 granted patents. About half of the issued patents were to companies overseas [5]. Of course, there are many more patent applications than issued patents because only about one-third of applications result in an issued patent.

Patent litigation is also important. Companies go to great lengths to protect their intellectual property and infringement penalties are high. In 2008, there were 2,909 patent suits filed [6].

In 1991, Eastman Kodak Co. was ordered to pay Polaroid Corp. $873,158,971 for infringing instant photography patents of Polaroid. The award was calculated based upon royalties, lost profits, and interest [7]. At the time, this was a record judgment. In 2010, Boston Scientific Corp. agreed to pay $1.725 billion to a unit of Johnson and Johnson Services, Inc. to settle long-running disputes between the companies over patents for coronary stents [8]. In August 2012, Dupont was found to infringe a Monsanto patent for seeds for herbicide-tolerant soybeans and ordered to pay $1 billion. That same month, in a separate case, Samsung was ordered to pay Apple $1.05 billion for infringing smart-phone patents. It is likely that these last two judgments will be appealed.

Nor is patent protection only applicable to large companies. In 1998, John Osher and three colleagues developed a spinning toothbrush, later named the SpinBrush®. They applied for and were granted a patent on the invention [9]. In 2001, they sold the technology to Procter and Gamble Co. for $165 million up-front and an agreement for future payments pegged to earnings. In 2002, they settled for a final additional payment of $310 million [10]. The Crest SpinBrush® became the nation's best-selling toothbrush, manual or electric.

Patents are useful because they give the assignee a competitive edge by preventing others from practicing the invention. Of course, the assignee can grant rights to others. By licensing the invention, the assignee has another source of revenue and many companies have made significant money by licensing technology.

If an invention is patentable, most elect to apply for a patent. Some will decide to forego patenting and keep the invention as a trade secret. This strategy is dangerous because the invention can later be considered to be abandoned and the inventor can lose his right to patent protection. Another risk is that a competitor will obtain patent protection for your trade secret and you will be barred from practicing the invention.

Issued patents are available on the USPTO website (http://www.uspto.gov) and patents issued after 1976 can be easily searched by inventor, assignee, or subject. Once on the website, click on "search for patents" and then "quick search." For example, a quick search of patents issued since 1976 reveals more than 40 issued patents with the inventor name, Einstein (not the Nobel Prize winning physicist), more than 3,800 patents with Pfizer as an assignee, and more than 500 with the concept, "chiral alcohol." Searching the patent literature is a valuable tool. It provides the reader with a synopsis of the art, with references, and best methods. Patents typically give experimental results. They are also a good way to understand the research direction and focus of a company because most companies patent their important results.

Often a patent search is an iterative process. For example, if I were interested in aziridine chemistry, I could search under the term aziridine. If I search on the U.S. Patent Office website as a quick search in patents issued since 1976, where the term aziridine is present in any field, the search shows more than 8,000 patents. This is too many and I need to refine the search. I can browse some of the 8,000 patents for ideas on other terms to use to limit my search. Another option is to refine the search to only show patents that have the term aziridine in a certain field such as the title, abstract, or experimental. If I refine it to include only those patents that have the term aziridine in the abstract, I now have 276 patents. If you repeat this, you may have slightly more because more patents may issue after the date of my search. I can browse these or refine further. If my interest is aziridine compounds used for control of diabetes, I can refine the search to include patents with aziridine in the abstract and the term diabetes anywhere in the patent. I now have a very manageable 14 patents. Within each of these patents, I can do further searching. For example, one of the 14 patents is U.S. Pat. No. 4,652,659, entitled, "Aziridine and phenethanolamine derivatives having antiobesity and anti-hyperglycemic activity." If I click on that patent, I get the text of the patent including a list of references. The references that are U.S. patents are hyperlinked and can be read merely by clicking on them. There is also a hyperlink entitled, "Referenced By." Clicking on that link reveals 18 subsequent U.S. patents that referenced U.S. Pat. No. 4,652,659. Viewing a patent on the U.S. Patent Office's website can be difficult especially the display of chemical structures; you may have to switch to image view. Another option is to find a pdf of a patent on another website such as Google. The pdf is easy to download and read and the chemical structures easily viewed.

The data can be searched by assignee. I might do this because I have a particular interest in a company. Maybe they are a competitor or I have an upcoming job interview with them or I just know that they are active in a research area of interest to me. As another example of a patent search, I know that McCormick and Company, Inc. makes spices, seasonings, and condiments. If I do a patent search with McCormick as the assignee, I see 122 patents. Most are issued to McCormick and Company, Inc. but some to McCormick and Munson Technologies, LLC, a Connecticut firm. If I browse the 122 patents, I see that many are design patents (designation begins with "D"). By reading others, I can get some sense of their technology. For example, U.S. Pat. No. 8,257,738, entitled "Encapsulation Compositions and Process for Preparing the Same" issued on September 4, 2012. By reading this and other McCormick patents, you can get some sense of their technology and research areas. You can also learn a little about the spice and seasoning business.

Sometimes searching by assignee can still be too broad. For example, Eisai Co., Ltd. is a pharmaceutical company headquartered in Tokyo. However, they employ more than 2,000 people in the United States. Searching under Eisai as the assignee shows over 900 patents. If I were interested in what type of work is performed at their Baltimore, MD facility, I could read the 900+ patents or I could refine my search to those where the assignee is Eisai and the inventor state is MD. If I do this, I find 26 patents and come to the conclusion that they are interested in cancer research at that location.

If you understand everything about patents, you might want to consult U.S. Pat. No. 4,608,967. The patent is for a self-congratulatory apparatus having a simulated human hand carried on a pivoting arm suspended from a shoulder-supported member. The hand is manually swingable into and out of contact with the user's back to give an amusing or an important pat-on-the-back.

If you get in too much of a celebratory mode, you may need to consult U.S. Pat. No. 85,659. The invention describes a new and improved antidote for drunkenness. The medicine is to be taken in doses of two to three table-spoonfuls three times per day. The list of 11 ingredients includes beef-gall, eel-skin, codfish, sweet milk, and cows' urine. This may be a case of the cure is worse than the disease.

QUESTIONS

1. In order to be eligible for a patent, an invention must satisfy three criteria. What are the three criteria?
2. On January 3, 2012, Madeline, working for the Cheery Chemical Company, has an idea for a new way to make cumene. Per company policy, that same day she writes the idea in her notebook, signs and dates

her notebook and has a coworker witness it. Madeline's boss, Dr. Sarah Noitall, a renowned synthetic chemist, tells Madeline that the idea is no good and cannot possibly work. Madeline gets upset and transfers to another group. On January 17, Abigail, Madeline's replacement who is unaware of Dr. Noitall's views, consults Madeline's notebook and begins working on the idea in the laboratory. Abigail is able to make cumene. Based upon what you have been told, choose the best of the following statements.

- **a.** There is no opportunity for a patent. Cumene is well known and has been made for years.
- **b.** It is likely to be patentable and Madeline is the sole inventor.
- **c.** It is likely to be patentable and Abigail is the sole inventor.
- **d.** It is likely to be patentable and Madeline and Abigail are coinventors.
- **e.** It is likely to be patentable and Madeline, Sarah, and Abigail are coinventors.

3. Eddie has an idea for a new process to make beer. His wife Anna, a physical therapist says, "Eddie, that's wonderful; I never would have thought of that process." His friends at the local tavern say the same thing. Eddie contracts with the Dewy, Cheathem and Howe law office and files a patent application with the USPTO. Eddie visits Budweiser, Sam Adams, and Coors headquarters. They confirm that his process is new and is likely to work. However, they all tell him that his process is an obvious variant of existing processes. It turns out that the distinction is that Eddie brews his beer two degrees hotter than the existing process. Is Eddie entitled to a patent?

- **a.** Yes, the process is new
- **b.** Yes, the process is useful
- **c.** No, the process isn't new
- **d.** No, the process is new but obvious
- **e.** Yes, the process is new, useful, and non-obvious.

4. It is okay and doesn't affect patentability to disclose an invention at a trade show more than one year before filing a patent application as long as the invention is not sold.

- **a.** True
- **b.** False

5. A U.S. Patent grants exclusive rights for how long?

- **a.** 10 years from the date of the invention
- **b.** 10 years from the date the patent application is filed

c. 10 years from the date the patent is issued

d. 20 years from the date the patent application is filed

e. for the lifetime of the inventor

6. Which of the following is a requirement for patentability in the U.S. Patent Office. (indicate all that apply)

a. must be commercially viable

b. must be useful

c. must be clever

d. must be non-obvious to one skilled in the art

7. Milton, MA, the birthplace of Buckminster Fuller, is a small town of about 27,000 outside of Boston. Do a quick search and see how many U.S. patents have issued since 1976 with at least one inventor from Milton, MA.

a. none

b. 27

c. almost 500

d. more than 700

e. more than 1,500

8. How many patents issued since 1976 are assigned to Pfizer where the inventor has the name Adams?

9. How many patents issued since 1976 have the term "fishing" and the inventor is from the city of Wilmington (note: searching like this includes inventors from any Wilmington such as Wilmington, NC, Wilmington, DE, Wilmington, MA, etcetera)?

a. none

b. 7

c. 17

d. more than 50

e. more than 100

10. How many patents, issued to Pfizer, have the term, "indole" in the title of the patent?

a. none

b. 7

c. almost 30

d. more than 50

e. more than 85

REFERENCES

1. Angela Murch. Intellectual Property Today 2013; 20(8):28.
2. 21 U.S.C. 355(b), (j), (l); 35 U.S.C. 156, 271, 282.
3. 35 U.S.C. 271(e)(1).
4. A Rufner, et al. U.S. Pat. No. 8,178,594. 2012.
5. Statistics from USPTO.
6. Intellectual Property Today 2009;16(8):27.
7. Polaroid Corp. v. Eastman Kodak Co.U.S. District Court District of Massachusetts 17 USPQ2d 1711 Decided January 11, 1991 No. 76-1634-MA.
8. Barry Meier. The New York Times. February 2, 2010, on page B3 of the New York edition.
9. Lawrence A Blaustein, John R Nottingham, John Osher, John W Spirk. U.S. Pat.No. 6,000,083. 1999.
10. Robert Berner. Business Week, Aug 12, 2002:58.

CHAPTER 5

Petrochemicals

5.1 CRUDE OIL

Crude oil comes from beneath the ground and is sold by the barrel. A barrel of oil is 42 gallons, so if oil costs $95 per barrel it costs $2.26 per gallon. Oil is traded on the open market and the price varies somewhat based upon the source of the oil. One common benchmark price is the price of Brent crude oil. Brent crude oil comes from the North Sea. Other oils, for example Dubai crude, may be priced somewhat higher or lower but generally the prices are similar. WTI oil stands for West Texas Intermediate. Crude oils can be light or heavy depending on their density. This should not be confused with the terms "lights" or "heavies", which refer to the fractions when oil is distilled in the refining process. The different oils have somewhat different compositions that can make them more or less desirable for gasoline production. Oil can vary in the composition of the distillates, having varying amounts of the different fractions, and can vary in the amount of sulfur. Oil that has less than 0.5% by weight sulfur is called "sweet". Oil with higher amounts of sulfur is called "sour."

For many different reasons, including demand, government regulation, and geopolitics, the price can vary remarkably. From 1988 to 2003 the price of a barrel of Brent crude varied from as low as about $10 per barrel to as high as about $35 per barrel. In 2004 the price started to climb and was at an all-time high of about $146 in the middle of 2008. From there, prices started to fall and were down to about $40 by the end of 2008. At that point they again started to rise. From January 2011 through January 2014, the price stayed between $90 and about $125 and averaged around $110 per barrel.

These same factors influence the spread among the different sources of oil. For example, in the last several years, the price spread between WTI and Brent crude has varied. In 2012, there was some concern about the European financial system and political unrest in the Mid-East. During this time,

Fundamentals of Industrial Chemistry: Pharmaceuticals, Polymers, and Business, First Edition. John A. Tyrell.

the purchase price of Brent oil was \$10–\$25 higher than WTI. However, in September of 2008, when the U.S. was undergoing the financial crisis induced by subprime mortgages, the spread spiked down with WTI trading at a purchase price of about \$22 higher than Brent. In January of 2014, Brent was trading about \$10 per barrel higher than WTI. The major use of oil is for gasoline and we can see the price fluctuations at the gas pump. However, many chemicals are based upon petroleum and their availability and cost is also influenced by the cost of oil.

The oil is refined, a process which converts the oil into usable products, mainly gasoline, but also several other chemical feedstocks. In the refining process, a continuous distillation is employed. Oil is continuously fed to a distillation column and different fractions are continuously removed from various heights in the column. Low boiling fractions are removed from the top, high boilers from the bottom and several other fractions from various places along the column. The lowest boiling fraction (light fraction) is a gas mixture and high boiling fractions are used as fuel oil or even for asphalt. An intermediate fraction is called naphtha. This can be further categorized as light naphtha or heavy naphtha. Naphtha is processed further by cracking or reforming to gasoline and other useful chemical feedstocks.

Cracking, as the name implies, breaks the molecules into smaller molecules, typically with more unsaturation. This chemical breakdown is done at high temperatures in the presence of a catalyst, hence the term, "catalytic cracking." One subset of catalytic cracking is the FCC (fluid catalytic cracking) process which uses a zeolite powder. Catalytic cracking in the presence of hydrogen is called hydrocracking.

We can start to get a sense of the chemistry by considering how n-heptane might be cracked. It is a radical mechanism and by heat, enough energy is supplied to break bonds.

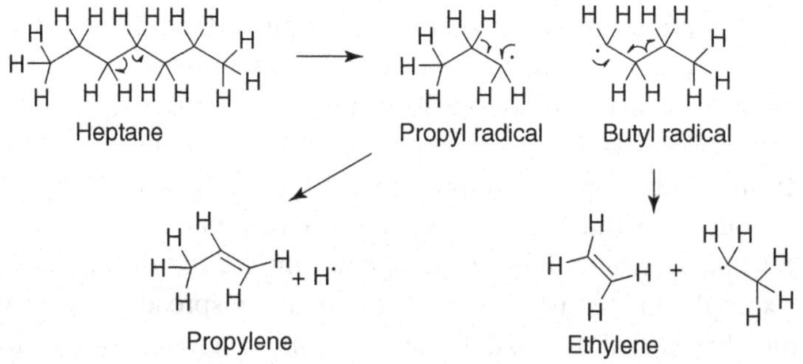

Heptane → Propyl radical + Butyl radical

Propyl radical → Propylene + H·

Butyl radical → Ethylene + ethyl radical

By breaking the C3-C4 bond, we create a propyl radical and a butyl radical. The propyl radical can decompose to propylene and a hydrogen radical. The butyl radical can form ethylene and an ethyl radical. The formed radicals can react further and from heptane, eventually several smaller molecules may be obtained. With a petroleum distillate, the chemistry is even more complex. The cracking feedstock might have several dozen different hydrocarbons. However, the chemistry is generally the same. Enough energy is supplied to cause homolytic bond cleavage. The resultant radicals react further. The cracked product consists essentially of smaller compounds and an increase in unsaturation. The cracked product is then separated by distillation. It is a source of a low boiling fraction containing hydrogen and methane gas. Subsequent fractions, in order of increasing boiling point are: ethylene; propylene; a C4 fraction; and then higher boiling fractions.

The C4 fraction will have 1,3-butadiene and other C4 hydrocarbons. Cracking conditions will have a major effect on the composition. The C4 components have similar boiling points so the 1,3-butadiene cannot be separated by a normal distillation process. Rather a technique known as extractive distillation is used. A solvent such as N-methylpyrrolidone (NMP) is added to decrease the volatility of the 1,3-butadiene. The high boiling fraction exits the bottom of the column and is the 1,3-butadiene and extracting solvent. This high boiling fraction is then separated by boiling and distillation to give 1,3-butadiene. The 1,3- butadiene is used to prepare many polymers including synthetic rubbers for a variety of uses including automotive tires. The raffinate, or material that remains after the extractive distillation, is the low boiling material that exits the top of the column.

N O
CH_3

1,3- butadiene NMP

The raffinate includes isobutene, also known as isobutylene, as the largest component followed by 1-butene and then 2- butene, both cis and trans isomers. Saturated butanes, n-butane and isobutane, are also present.

isobutene 1-butene trans-2-butene cis-2-butene n-butane isobutane

Another process is reforming. Reforming is done with petroleum fractions that have an octane rating too low to be suitable for gasoline. Gasoline octane rating is an indication of the propensity of the gasoline/air mixture to prematurely self-ignite in the combustion chamber of the engine. This premature ignition exhibits itself as engine knocking. Isooctane, 2,2,4-trimethylpentane, has an octane rating of 100 and n-heptane has an octane rating of 0. Gasoline with an octane rating of 90 would have similar combustion behavior to a 90:10 ratio of isooctane : n-heptane. Recognize that something can have an octane rating of 90 or even 100 and contain no octane or isooctane. The rating has to do with the combustion properties and not how much octane or isooctane is in the fuel. Octane ratings generally increase with branching. In reforming, the petroleum distillate is heated in the presence of an isomerization catalyst, such as halogen-treated alumina or non-acidic zeolite L, and a hydrogenation/dehydrogenation metal, e.g., a Group 8, 9, or 10 metal such as platinum. This forms branched isomers and cyclic compounds. Distillation of the reformate gives a hydrocarbon stream of increased octane level. The distillation is also a source of BTX, a mixture of benzene, toluene, and xylene

There are several separation techniques used to isolate the various components of the petroleum product stream. The most common is distillation. Distillation depends upon a mixture having a different composition in the vapor phase from the composition of the liquid phase. Consider a simple distillation of a 50-50 molar ratio of a two-component mixture. When the mixture is heated to the boiling point, typically the vapor is enriched in the lower boiling component. The enrichment will vary based upon the difference in boiling points. If the vapor has an 80:20 molar ratio of low boiler to high boiler, condensation of that vapor will result in a fraction that is 80% (molar basis) of the lower boiling component. If that were distilled again, the collected vapor may be 95% of the low boiler. Repeated distillation can give material which is an even purer low boiling fraction. At the same time that the low boiler is obtained by the collection of the vapor phase, the high boiler is obtained by collection of the liquid phase. Rather than doing several of these simple distillations over and over again, a packed column is used. In the column, the vapor condenses on the packing material and then some of it is returned as liquid back down the column and some of it revaporizes; essentially it is redistilled. Each condensation and revaporization is like a simple distillation. The number of distillations within a column is called the number of plates. A column is said to have a certain number of theoretical plates. Depending on the surface area of the packing material, the height of the column, and the return ratio of the condensed liquid back down the column, the column can be the same as doing many simple distillations. Industrial distillation columns can be many stories high and can have over 100 theoretical plates. Because of their size and height, they can often be seen from a distance. These distillation processes are typically run in a continuous manner with feedstock constantly entering the column and fractions continuously being removed from various points along the column.

Fractional distillation by itself is not always suitable for isolation of the desired component. For example in the separation of BTX, alkanes such as cyclohexane, n-heptane and others can codistill with benzene and toluene.

Among the C8 fraction, ethylbenzene (136.2 °C), p-xylene (138.3 °C), m-xylene (139.1 °C) and o-xylene (144.4 °C), have similar boiling points. Other separation processes are often used [1].

One of these is azeotropic distillation. In an azeotrope, the composition of the liquid and the vapor is the same. If a liquid is at the azeotropic composition, it cannot be purified by distillation. In an ideal mixture, the tendency for a component to vaporize is the same in the mixture as in the pure component. However in non-ideal mixtures, the intermolecular forces between the components are different than in the pure liquid. This happens with ethanol and water. If you were to take a 50% (by mass) ethanol mixture and boil it, the composition of the vapor would be enriched in the lower boiling ethanol. Condensation of the vapor would result in a composition greater than 50% ethanol. Redistillation might increase the ethanol composition further. However, when we reach 95.6% ethanol, the composition of the vapor is the same as the composition of the liquid. Therefore, distillation of 95.6% ethanol does not result in an increase in the percentage of the lower boiling ethanol in the condensed vapor. Rather, the condensed vapor is 95.6% ethanol. This mixture is known as a constant boiling mixture or an azeotropic mixture. The boiling point of this particular mixture is lower than that of either pure water or pure ethanol, so this is called a low boiling azeotrope. For other mixtures, we can have high boiling azeotropes, where the boiling point of the mixture is higher than either of the components. To this point, we have discussed two-component or binary azeotropes, but recognize that they can be more complicated and can have three (ternary azeotropes) or even more components.

One separation process takes advantage of azeotropes. An auxiliary agent such as acetone or methanol is added to a mixture of alkanes, cycloalkanes, and predominantly aromatics. The agent forms a low boiling azeotrope with the alkanes and cycloalkanes thereby enabling their removal from the aromatics. The remaining agent is then often extracted from the aromatic high boiling stream and recycled.

Another process that is sometimes used is extractive distillation [2, 3]. The concept of extractive distillation is that a non-volatile polar solvent such as sulfolane is added, which has a different effect on the volatility of the components of the hydrocarbon mixture. Typically, the solvent decreases the volatility of the aromatic compounds, making them easier to separate from the aliphatics. The solvent is continuously added near the top of the extractive distillation column. Aliphatics are removed overhead and solvent and aromatics from the bottom of the column. The aromatics can be separated from the extractive solvent by distillation in a solvent recovery column. The boiling point of sulfolane (285 °C) is significantly higher than even o-xylene (144.4 °C) enabling the isolation of the aromatic stream. Optionally, this can be further purified by a water wash.

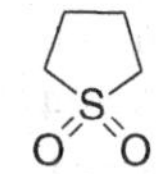

Sulfolane

If the aromatic content is low, the aromatics can be extracted from the hydrocarbon mixture by a liquid–liquid extraction. This is the same principle as performed in a small scale in the laboratory with a separatory funnel with two immiscible liquid layers. On the industrial scale, it is done continuously in a counter-current extraction column.

Crystallization is used to separate p-xylene from its C8 isomers. The melting point of p-xylene is 13 °C which is significantly higher than the other xylenes or ethylbenzene. Therefore, p-xylene can be selectively crystallized by cooling a C8 mixture.

Another method to separate p-xylene relies on the difference in shape of the isomers. The para isomer can be selectively adsorbed into pores of certain zeolites. It is then rinsed out of the zeolite bed with a desorbing solvent that can be easily separated by distillation.

5.2 COAL, NATURAL GAS AND SHALE OIL

In addition to oil, coal and natural gas are other fossil fuels that are used for energy and conversion to higher value organic chemicals. Coal has been used as a fuel source since the cave man. Historically, and to some extent even today, coal has been used to produce chemicals. However, most coal is used by power plants to produce electricity. In the United States, this represents over 90% of consumption with the other major use as a fuel for heat and power. About a billion tons of coal are mined each year in the United States. Worldwide production is about 8 billion tons.

Natural gas is predominantly methane. The composition varies but it can also include significant amounts of ethane, propane, butane, and isobutane. Other gases such as carbon dioxide, nitrogen, hydrogen sulfide, and helium can also be present. In this state, with the heavier hydrocarbons, it is sometimes referred to as "wet gas." Natural gas is available from several sources including some derived from oil and some from coal beds. Tight gas refers to natural gas located in impermeable hard rock and shale gas is natural gas entrapped in shale deposits. The U.S. produced about 23 trillion cubic feet of natural gas in 2011 and this is likely to increase. One reason is the increase in production from shale deposits. Shales are fine-grained sedimentary rocks that can hold petroleum or natural gas. Historically they have been drilled by vertical wells and infusing chemically treated water and sand to break up the rock in a process called "fracking." When the shale is broken up, entrapped natural gas is released. Drilling techniques have improved to allow turning the vertical shaft as much as 90 degrees to then allow horizontal drilling. This enables the mining of a much larger area with a single well. The discovery of large shale deposits, improvements in fracking technology, and the ability to drill horizontally have remarkably improved the economics of natural gas. The technique is not without controversy. There are environmental concerns about the injection of large volumes of the fracking water and how they may affect drinking water supplies. There can also be issues with nearby drinking wells and contamination with methane. Natural gas is purified by a refrigeration and distillation process. This pressurized liquid purified product is mainly methane and is called liquefied natural gas (LNG). The distinctive smell of LNG is caused by low levels of an additive, a mixture of mercaptans, which gives LNG the distinctive rotten-egg smell for safety and leak detection. LNG is mainly used as a fuel source, much of it in electrical power plants. The removed higher hydrocarbons such as ethane or propane can also be used as fuels or can be converted to feedstocks such as ethylene or propylene.

Oil shale rock is another source of oil and energy. This rock can be mined and then heated to produce oil as well as steam for energy. Oil shale is sedimentary rock that contains organic matter. The amount of organic matter varies and is as high as 50% in some high grade deposits, although 5 to 25% is more common [4]. Compared with fracking, which involves deep wells and reserves, oil shale is located near the surface.

5.3 ETHYLENE

The world's 140 million metric tons of annual ethylene capacity almost exclusively employs steam cracking of hydrocarbon feedstocks [5]. The majority of the feedstocks come from petroleum refining, such as by cracking of naphtha, but some producers use liquefied natural gas as a feedstock. In Brazil, where sugar cane is plentiful, Braskem has built a 200,000 metric ton per year ethylene plant based upon the dehydration of sugar-derived ethanol [6]. In the United States, natural gas liquids, a mixture of ethane, propane, butane, and other hydrocarbons, are available from shale deposits. The ethane is separated and cracked to make ethylene. Depending on the cost of oil and natural gas, this can be an economic advantage. In 2012, about 70% of United States ethylene production was from ethane [7].

Ethylene is the largest volume organic chemical produced in the world. Much of it is used captively for conversion into other products. The largest use is for the polymerization of ethylene to make polyethylene. Polyethylene has many desirable properties including low cost, and therefore each year it is the largest volume polymer produced. Other ethylene derivatives, in approximate order of volume of ethylene consumed, are ethylene dichloride, ethylene oxide, and ethylbenzene.

Ethylene dichloride (EDC) is made by the direct chlorination of ethylene [8] in the presence of a catalyst such as iron (III) chloride or copper (II) chloride. EDC is also made by oxychlorination, a process where ethylene reacts with HCl and oxygen in the presence of a catalyst, often copper (II) chloride [9, 10]. The EDC is then converted to vinyl chloride, commonly called vinyl chloride monomer (VCM) by a dehydrohalogenation reaction. The VCM is then polymerized to polyvinyl chloride (PVC).

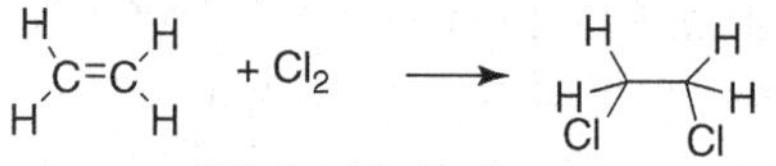

$$H_2C{=}CH_2 + Cl_2 \longrightarrow H_2ClC{-}CClH_2$$

Direct chlorination

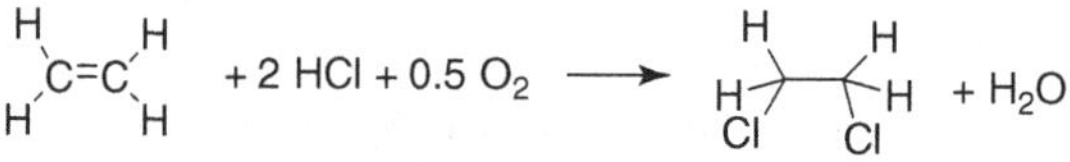

$$H_2C{=}CH_2 + 2\ HCl + 0.5\ O_2 \longrightarrow H_2ClC{-}CClH_2 + H_2O$$

Oxychlorination

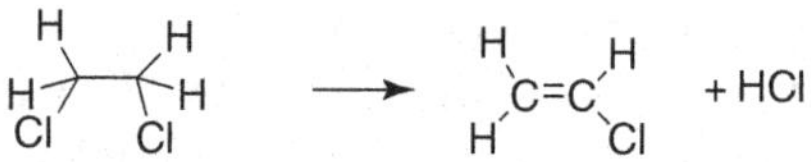

$$\underset{\text{EDC}}{H_2ClC{-}CClH_2} \longrightarrow \underset{\text{VCM}}{H_2C{=}CHCl} + HCl$$

Dehydrohalogenation

The dehydrohalogenation of EDC to VCM is an endothermic reaction. It is sometimes referred to as a thermal cracking reaction. The heat can be supplied by the reactions that produce EDC. Both the direct chlorination and the oxychlorination reactions are exothermic.

The production of VCM generates an equivalent of HCl. The oxychlorination process provides an outlet for this HCl byproduct. In any large-scale process it is critical to have an outlet for all byproducts. Although it is performed in different reactors, the overall conversion of ethylene to VCM can be written.

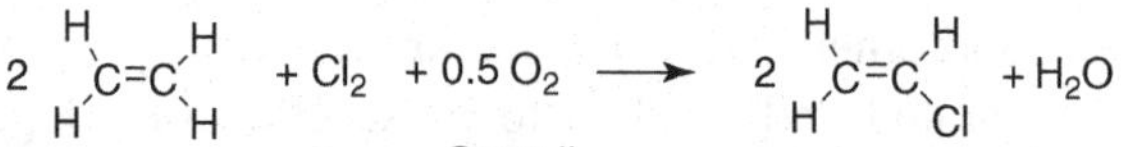

$$2\ H_2C{=}CH_2 + Cl_2 + 0.5\ O_2 \longrightarrow 2\ H_2C{=}CHCl + H_2O$$

Overall

An alternative process to VCM is from calcium carbide. This process is used in China [11]. Calcium carbide, CaC_2, is made by heating lime with coal-derived coke. After treatment with water, acetylene is formed. The acetylene is then reacted with HCl to produce VCM.

$$CaO + 3\,C \longrightarrow \underset{\text{Calcium carbide}}{CaC_2} + CO$$

$$CaC_2 + H_2O \longrightarrow \underset{\text{Acetylene}}{H{-}C{\equiv}C{-}H} + CaO$$

$$H{-}C{\equiv}C{-}H + HCl \xrightarrow{HgCl_2\ \text{catalyst}} \text{VCM}$$

Ethylene oxide is an epoxide made by the oxidation of ethylene in the presence of a catalyst, often a silver catalyst [12]. Ethylene and air are passed over a catalyst, typically at pressures of 10–30 bar and temperatures of 200–300 °C. The reaction is exothermic and the exotherm must be controlled to prevent runaway reaction and further oxidation to carbon dioxide. A typical reactor consists of large bundles of several thousand tubes that are packed with catalyst. A coolant surrounds the reactor tubes, removing the reaction heat and permitting temperature control [13]. Ethylene oxide has many uses including the production of various ethers, alcohols and ethanolamines, but the largest use is the production of ethylene glycol. Ethylene oxide is converted to ethylene glycol by reacting the ethylene oxide with an excess of water in the presence of an acid catalyst. The acid catalyst is often a solid, such as polymer-bound sulfonic acids. This enables the acid to be readily separated from the reaction mixture. Although only one equivalent of water is required by theory, a large excess, perhaps 10:1 to 20:1 water: ethylene oxide on a molar basis, is used because of the competing reaction between unreacted ethylene oxide and ethylene glycol to form diethylene glycol.

$$\underset{\text{Ethylene}}{H_2C{=}CH_2} \xrightarrow[\text{Catalyst}]{O_2} \text{Ethylene oxide}$$

$$\text{Ethylene oxide} + H_2O \xrightarrow{H^+} HO{-}CH_2CH_2{-}O^{\oplus}H_2 \xrightarrow{-H^+} \underset{\text{Ethylene glycol}}{HO{-}CH_2CH_2{-}OH}$$

$$\text{Ethylene oxide} + \underset{\text{Ethylene glycol}}{HOCH_2CH_2OH} \xrightarrow{H^+} HOCH_2CH_2{-}O^{\oplus}H{-}CH_2CH_2OH \xrightarrow{-H^+} \underset{\text{Diethylene glycol}}{HOCH_2CH_2OCH_2CH_2OH}$$

Most ethylene glycol is used to make polyethylene terephthalate, the largest volume polyester. Diethylene glycol contaminating the ethylene glycol will also polymerize and result in a polyester with lower melting point and poorer physical properties. The other major use of ethylene glycol is antifreeze. When mixed with water, ethylene glycol lowers the melting point of water. This is because of the freezing point depression which occurs when a solute is dissolved in a solvent. This is the same phenomenon that explains why salt water has a lower freezing point than pure water. The sodium and chloride ions are the solute and lower the freezing point of water, the solvent. In the case of ethylene glycol, it is soluble in water and lowers the freezing point of water. When temperatures are cold, it prevents the water in automobile radiators from freezing. When water freezes, it expands and can crack and break parts of the engine. Other solutes lower the freezing point, but have disadvantages. Methanol or ethanol is too volatile; sodium chloride is too corrosive. Ethylene glycol is good for this purpose. A little diethylene glycol contaminating the ethylene glycol does not prevent it from being useful for antifreeze, but large amounts of diethylene glycol lower the effectiveness. This is because the amount of the freezing point depression is based on the number of moles of solute. Pure ethylene glycol has a greater number of moles in a gallon than a gallon of an 80:20 molar ratio mixture of ethylene glycol:diethylene glycol.

Ethylbenzene is also made from ethylene. This is a Friedel–Crafts alkylation reaction, a type of electrophilic aromatic substitution.

Ethylbenzene can also be isolated from the C8 stream of the BTX petroleum fraction. Ethylbenzene is dehydrogenated to make styrene, which in turn is used to make polystyrene. High temperatures, about 600 °C, and a catalyst, often an iron oxide, are used to do the dehydrogenation.

5.4 PROPYLENE

Like ethylene, propylene is produced by the cracking of hydrocarbons. The major use of propylene is to make polypropylene but it is also a feedstock to manufacture other industrial chemicals, including propylene glycol, acrylic acid, propylene oxide, cumene, and isopropyl alcohol. Propylene oxide is made by the oxidation of propylene, and hydrolysis of propylene oxide gives propylene glycol. The alkylation of benzene with propylene gives isopropyl benzene, more commonly called cumene. Note that the product is not n-propyl benzene. This is because the intermediate is the more stable secondary carbocation which results in isopropyl benzene.

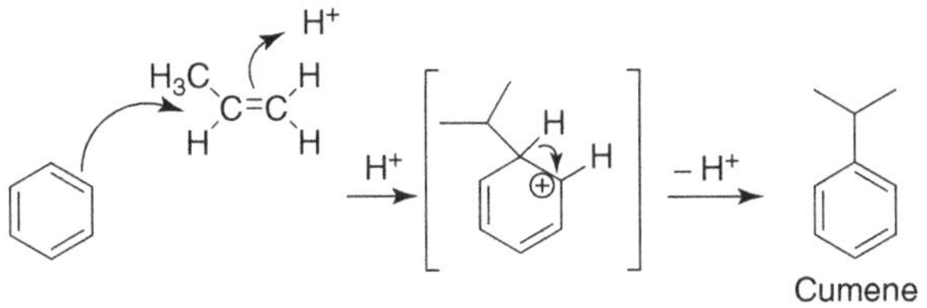

Cumene is converted to phenol and acetone. The reaction involves the oxidation of cumene to cumene hydroperoxide. This can be done in a radical process by passing air through cumene. In a second step, the cumene hydroperoxide is then treated with acidic water to form phenol and acetone in a 1:1 molar ratio.

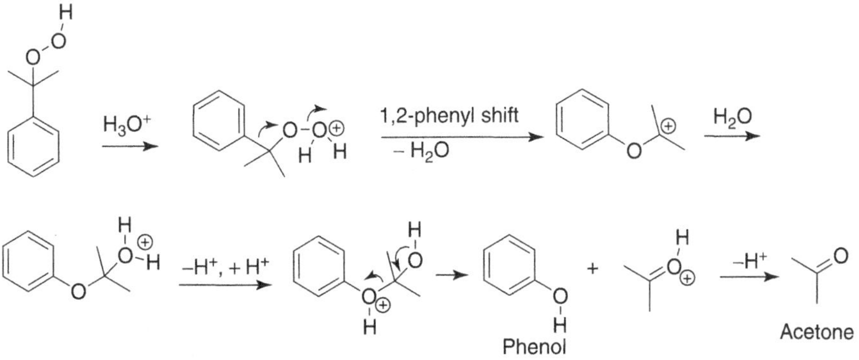

Phenol and acetone can be reacted together in a 2:1 molar ratio to make bisphenol-A (BPA). This is an electrophilic aromatic substitution reaction with the electron-rich phenol ring and acetone as the electrophile.

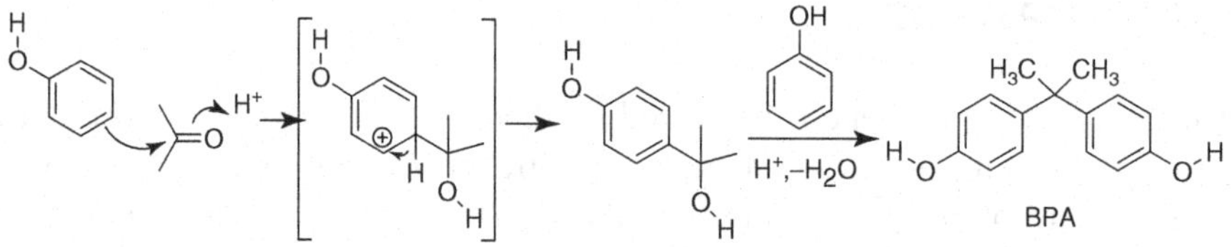

Two molecules of phenol are used for each mole of acetone. Because phenol and acetone are produced in a 1:1 molar ratio and consumed in a 2:1 molar ratio, there is an excess of acetone. Therefore acetone is often a glut on the market and is a generally inexpensive solvent. The BPA is used to make various polymers, including polycarbonate, many epoxy resins, and some phenolic resins. BPA is used in many polymers that have food contact and BPA can be present in minute amounts. Examples of these applications include some plastic containers, bottles, and can linings. Therefore, there is some concern about possible health effects. It continues to be a matter of study, but one review of the research suggests that current exposure levels are safe [14]. Phenol is also used to make phenol resins and as a feedstock in the manufacture of caprolactam, the monomer for nylon 6.

Acrylic acid represents a 4.5 million metric ton per year global market with BASF, Dow, and Arkema being major manufacturers [15]. Acrylic acid is manufactured by the oxidation of propylene. The reaction is a two-step process with oxidation to acrolein followed by further oxidation of the aldehyde to the carboxylic acid. Acrolein is a raw material for cosmetics, flavors, and pharmaceuticals. The major use of acrylic acid is as a monomer to make acrylic acid polymers and conversion to acrylate esters which are also used to make polymers.

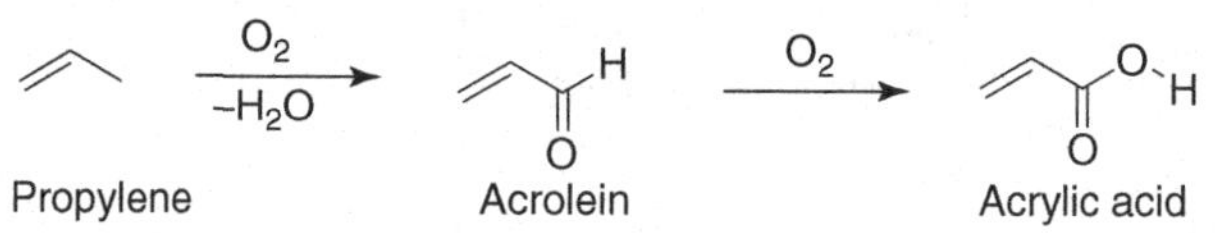

Although most acrylic acid is produced from propylene, there are other routes including processes that begin with ethylene oxide or acetylene. One process being developed independently by Arkema and Nippon Shokubai [16] is based upon glycerol as a starting material. Glycerol (also known as glycerin or glycerine) is a byproduct of biodiesel formation. A vegetable oil is transesterified with methanol to give the methyl ester, used as biodiesel, and glycerol. The R groups vary with different vegetable oils but are typically 11 to 17 carbons in length. The glycerol can be converted to biobased acrylic acid either directly or via acrolein.

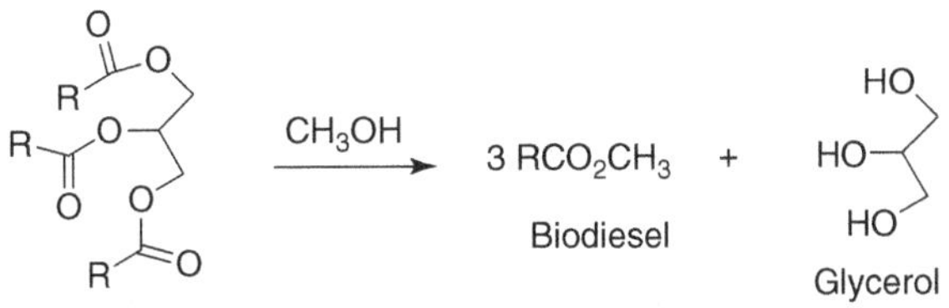

CH_3OH → $3\ RCO_2CH_3$ + Glycerol

Vegetable oil
R = Long chain hydrocarbon

Biodiesel

Glycerol

5.5 BTX

BTX is a mixture of benzene, toluene, and xylene. It comes from petroleum refining. Depending on the source, the composition will vary. For example, BTX from the cracking of naphtha is typically richer in benzene than BTX from a gasoline reforming process. The boiling points are sufficiently different that they can be separated. In approximate order of importance, benzene is used to make ethyl benzene, cumene, cyclohexane, and nitrobenzene as well as many other chemicals. Toluene is used as a solvent and to make toluene diisocyanate, a monomer used to make polyurethane. Toluene can also be disproportionated to xylene and benzene.

Of the xylenes, p-xylene has the largest volume use. It is oxidized to terephthalic acid, which is used as a monomer to make polyesters. Ortho xylene is oxidized to phthalic anhydride. Phthalic anhydride can be reacted with long-chain alcohols to make plasticizers to impart flexibility to PVC resin. Phthalic anhydride is also used to make unsaturated polyester resins.

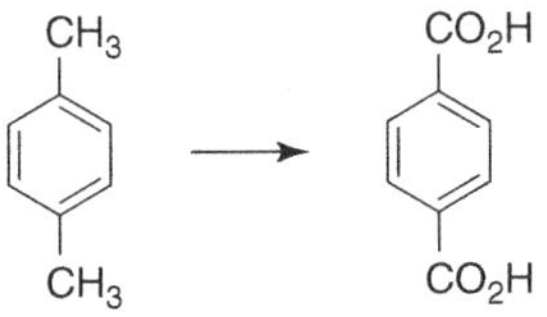

p-xylene → Terephthalic acid

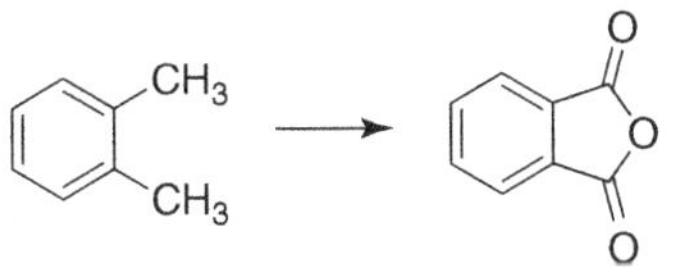

o-xylene → Phthalic anhydride

QUESTIONS

1. Which of the following is always true about a two-component mixture that forms an azeotrope?
   a. the boiling point of the azeotrope is higher than the boiling point of the lower boiling component
   b. the boiling point of the azeotrope is lower than the boiling point of the lower boiling component
   c. at the azeotrope, the liquid and vapor have the same composition
   d. separation of the two components by distillation requires a longer fractionating column.

2. The boiling point of p-xylene is very close to that of the other C8 isomers, making separation by distillation difficult. Name two ways that p-xylene is commercially separated from the C8 stream.

3. A plant that produces 116 tons per year of acetone coproduces ________ tons of ________?

4. Ethylene glycol is made from ethylene oxide and a large excess of water. Draw the structure for the major by-product if one equivalent (rather than an excess) of water is used.

5. Consider the following two reactions to make EDC.

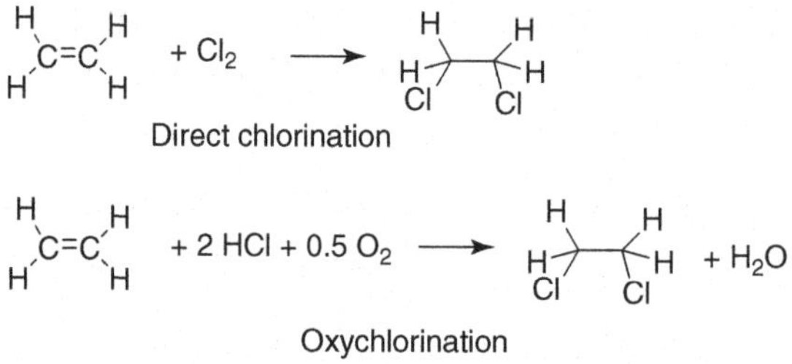

$$H_2C{=}CH_2 + Cl_2 \longrightarrow ClH_2C{-}CH_2Cl$$

Direct chlorination

$$H_2C{=}CH_2 + 2\ HCl + 0.5\ O_2 \longrightarrow ClH_2C{-}CH_2Cl + H_2O$$

Oxychlorination

   a. Most EDC is made by direct chlorination
   b. Most EDC is made by oxychlorination.
   c. Most EDC is made by using both direct chlorination and oxychlorination in about a 1:1 ratio.
   d. Most EDC is made by using both direct chlorination and oxychlorination in about a 2:1 ratio.
   e. Neither of these reactions is used today because of environmental concerns.

6. The price of a barrel of oil can fluctuate remarkably. Although this price affects everything to some extent due to transportation and fuel costs, etc, some industries which rely on petroleum products as raw materials are greatly affected by the price of a barrel of oil. From the following list, which are greatly affected (depend on petroleum products as starting materials):

NaOH

sulfuric acid

vinyl chloride

ammonia

xylene

REFERENCES

1. K Weissermel, H-J Arpe. *Industrial Organic Chemistry*. 3rd ed. New York: VCH Publishers, Inc.; 1997:317.
2. L Tian, et al. U.S. Pat. No. 7,078,580. 2006.
3. JA Vidueira. U.S. Pat. No. 5,310,480. 1994.
4. Eric Niiler. Chemical and Engineering News 2013; 91(30):25.
5. Alexander Tullo. Chemical and Engineering News 2011; 89(3):20.
6. Michael McCoy. Chemical and Engineering News 2010; 88(33):25.
7. Alexander Tullo. Chemical and Engineering News 2012; 90(10):11.
8. Joseph A Cowfer. U.S. Pat. No. 4,760,207. 1988.
9. Albert Kister, U.S. Pat. No. 3,892, 816. 1975.
10. Keith S Kramer, Joseph A Cowfer. U.S. Pat. No. 7,585,806. 2009.
11. Jean-Francois Tremblay. Chemical and Engineering News 2010; 88(3):18.
12. Masaharu Kiriki, Hitosh Takada. U.S. Pat. No. 6,114,553. 2000.
13. DM Rekers, AA Smaardijk. Eur. Pat. No. 2121646. 2009.
14. Britt E Erickson. Chemical and Engineering News 2011; 89(17):28.
15. Marc Reisch. Chemical and Engineering News 2010; 88(21):20.
16. Marc Reisch. Chemical and Engineering News 2010; 88(21):20.

CHAPTER 6

Business Considerations

6.1 INTRODUCTION

Whether one is working in the chemical industry or collaborating with members of the chemical industry, there are factors beyond pure science that come in to play and are factors in whether projects live or die and how they are done. It is helpful to have some knowledge of these considerations. Even if you do not use the various techniques, it is worthwhile to understand them because decision-makers in your company are likely using them.

In any organization, there is a desire to improve quality, whether it is the quality of a product or the quality of processes within the company. Lack of quality costs money and puts companies out of business. Quality is dynamic so there is always room for improvement. Top companies have a focus on quality. There are several techniques used to improve quality; one common one is Six Sigma.

All projects within a company are not successful but all projects require time, resources, and money. The earlier "bad" projects are abandoned, the less investment in bad projects and the greater is the ability to put resources on other projects with a greater chance for success. There are different ways to choose which projects to back; often it is merely a judgment by someone in upper management. There are more refined techniques to improve this process and Stage-Gate™ is a formalized method to judge projects.

Most people need to work collaboratively. Projects are usually too large for a single person to accomplish. Communication and trust are important tools to help collaboration. The way a company is organized can facilitate or detract from collaboration. There are different types of organization and some understanding of these is warranted. When several people work on the same project, it is important for each to understand their role in project completion. This can be difficult, especially regarding the overall timing of

---

Fundamentals of Industrial Chemistry: Pharmaceuticals, Polymers, and Business, First Edition. John A. Tyrell.

projects and to understand the effect that delays in one piece may have on the overall project timing. One technique to track, understand and communicate the various aspects of a project is through the use of Gantt charts.

With any product that is sold, profitability is dependent upon costs. When the product is a chemical, the major cost is usually the raw materials needed to produce that chemical. Early in any project, raw material costs should be calculated. If they are too high, the project will not be successful. Also when a chemical is first made in larger scale, there are many factors that need to be considered. By thinking of them in advance, there is a greater chance for success.

This chapter tries to introduce the reader to some of these concepts. By thinking about these things, better decisions will be made and there will be greater productivity.

6.2 SIX SIGMA

Six Sigma is a process to improve performance. There are thick books on the topic and entire courses devoted to training someone on the topic so what follows is just a brief synopsis of the technique. The name derives from the Greek letter "σ", or sigma, which is used as an abbreviation for standard deviation. The standard deviation is the square root of the variance. If the data fits a bell-shaped normal distribution curve, 68% of the numbers will fall within one standard deviation of the mean, 95% within two standard deviations, and 99.7% within three standard deviations. Six Sigma has a focus on decreasing the number of defects. A defect is anything out of specification.

Recognize that a single unit can have more than one defect. A polymer sample might be defective (or failing a specification) in both color and in impact. We can measure the defects per unit by dividing the number of defects by the number of units. Imagine two processes of varying complexity. In one process, steel rods are manufactured. The only customer specification is that the length must be within five mm of one meter. They average one defect per 1,000 rods. They are averaging 999 defect-free rods per 1,000. In another process, smart phones are manufactured. For these phones, there are 25 specifications. This manufacturer also averages 1 defect per 1,000 units (phones). However, they have noticed that a defective phone typically has several defects. They average 999.8 defect-free phones per 1,000 phones. Which process has better quality? They both average one defect per 1,000 units, but we can argue that the process to make the phones has better quality.

If we think about each opportunity for a defect to occur, we can measure defects per opportunity and not per unit. In the case of the steel rods, we would have a defect rate of one per 1,000 opportunities or 1,000 defects per million opportunities. For the phones, we have one defect per 25,000 opportunities or 40 defects per million opportunities. This value is probably more representative of the quality of the manufacturing processes. We can account for the complexity of the products and the number of specifications by considering the value of defects per million opportunities or DPMO.

There is typically less variation in a short period of time than in a longer period of time. To account for this longer term process variability, an empirically-based 1.5 sigma shift is introduced into the calculation. Based upon this idea, a process that fits six sigma between the process mean and the nearest specification limit in a short-term study will in the long term fit only 4.5 sigma. A process that is termed a six sigma process is more strictly speaking a 4.5 sigma process. There are charts [1] that correlate DPMO with the capability of the process. A process that has 3.4 DPMO is a six sigma process. In contrast, a three sigma process has 66,800 DPMO. The goal of a six sigma program is to reduce process defects to a level of 3.4 DPMO. At first, this might seem too intense. Why not target four sigma? A process running at four sigma has 99.4% of its output defect free. That may sound good, but not when you focus on DPMO. The DPMO of a four sigma process is 6,210. If operating at four sigma, every hour the post office would lose 20,000 pieces of mail and every week, there would be 5,000 surgical operations that go wrong in some way [2].

Six Sigma is a method to reduce variation and therefore the number of defects to a level of 3.4 DPMO. Six Sigma uses statistical tools to identify the important factors that can improve the quality of the processes. There are five important phases [3]: 1. Define the projects, goals, and deliverables to customers; 2. Measure the current performance of the process; 3. Analyze and determine the root cause of the defects; 4. Improve the process to eliminate defects; and 5. Control the performance of the process. These phases go by the acronym DMAIC. In the Define phase [4], the goals of the improvement activity are determined. These could come from internal or external customers or could be strategic objectives of the organization or improved throughput and reduced defects. In the Measure phase, valid and reliable metrics are established to monitor progress toward the goal. In the Analyze phase, ways to eliminate the gap between current performance and desired goal are explored using statistical analysis. In the Improve phase, new ways to do things are implemented and statistical methods are used to validate the improvement. The statistical tools are then used in the Control phase to monitor the stability of the new system. Policies, procedures, and documentation are implemented to enforce the new process.

Implementation of Six Sigma involves people throughout the organization and the training of many people. Sometimes their roles are given names such as "Black Belts" or "Green Belts", based upon the amount of training and the role they are expected to play. For example, Six Sigma Black Belts [5] should be technically oriented highly regarded individuals who work to extract knowledge from the organization. They might receive three to six weeks of training. Green Belts are project leaders capable of forming and facilitating teams and might receive a week of training. They are often advised by Black Belts.

As the DMAIC process is followed, it is important that a phase be completed before the next phase commences. It does not make sense to embark upon the measurement phase prior to having a clear definition of what is important. This can be done informally or by a formal "gate" mechanism where the project cannot proceed to the next phase until it undergoes a gate review.

Lean is an effort to improve efficiency by removing tasks that do not add value. When it was implemented at GE, it was called "work out" because the focus was to get rid of unnecessary work. Even seemingly little things like filling out time cards or writing reports can take time away from efforts that add value to the customer. By eliminating unimportant tasks, process steps, reports, reviews and other practices, the organization can focus on work that adds value, be more efficient and be more agile. Lean requires you to know what is important and eliminate the unimportant. The implementation of Lean delivers speed and low cost. The use of Six Sigma techniques in a focused effort to reduce wasted efforts is called Lean Six Sigma.

These techniques can be used throughout an organization or in smaller pieces such as a particular chemical process. For example Lean Six Sigma has been used to improve a selenium analysis [6] and showed the importance of improving the digestion of the sample.

6.3 STAGE-GATE™

There is never enough time and resources to pursue everything to completion. Also why would you want to? Most projects fail. In the chapter on the pharmaceutical industry, we discuss the failure rates. For drug candidates that enter Phase 1 testing, there is less than a 12% success rate. Even those entering Phase 3, have a success rate around 50% [7]. Think about what this means. After 10–15 years and perhaps $1 billion in costs, you still fail half the time. Failure is not peculiar to the pharmaceutical industry, but is common in all new product research. Yet we need to introduce new products to remain competitive. The key is being prudent about which projects to pursue and which ones to shut down. As an analogy, if you are driving or hiking somewhere new to you and take a wrong turn, the sooner you recognize it and turn back, the less time and energy you have lost and the less time to backtrack. Similarly, the sooner we recognize a fatal flaw with a project and kill it, the sooner we can shift resources – people, money, equipment, etc. – to another project with a better chance for success. However, these decisions are not easily made. You run the risk of killing a good project.

Thomas Edison said, "Many of life's failures are men who did not realize how close they were to success when they gave up." Making project decisions becomes a balancing act but an important one to success. Lipitor® is one of the more important drugs in recent history. Millions of people take it to control cholesterol. It went off patent Nov. 30, 2011 and is now available in generic form. In 2010, it generated more than $10 billion in sales. Lipitor® was invented at Parke-Davis, since acquired by Pfizer. In 1990, the Parke-Davis review board met to decide whether to continue the project and begin Phase 1 studies. Researchers had worked on the project for eight years, but at best it would be the fifth statin to reach the market and animal studies did not distinguish it from other statins. Although the sentiment was to kill the project, Dr. Roger Newton was able to convince the review board to continue with the project [8]. The Lipitor® story is an example of when it was desirable to continue with the project. The decision had a major humanitarian and financial impact.

Because of the importance of these decisions many companies have elected to have a formalized process to review projects and decide which projects should be supported and which should be discontinued. Without a rigorous method, the wrong projects often get selected. Instead of decisions being based upon facts and objective criteria, they are often based upon company politics and emotion with the end result that too many fail [9]. One of these formalized processes was introduced by Robert Cooper and is called the Stage-Gate™ process. It is described in his popular book first introduced in 1998 and now in its fourth edition [10].

There are typically about six stages or phases from conception to commercialization. For each stage, there are defined learnings or tasks to be done. A group meets and reviews the work done in the stage and decides whether there is sufficient probability of success to warrant moving to the next stage. This is called passing through a gate and the members of the group are called gate-keepers because they decide which projects pass through the gate. The project can be killed, sent back for more investigation or passed along to the next stage. The composition of the gate-keepers might change as a project moves through the stages. At an early stage, perhaps a literature search of the chemistry and a preliminary market study is done and the gate keepers might be an experienced scientist and someone from new business development. At a later stage, as more investment is needed, representatives from research and development (R&D), sales, marketing, manufacturing, and upper management are all likely to contribute and to take part in the decision whether to move the project into the next stage, with the ultimate stage being commercialization. The overall goal is to pick the best projects based upon defined objective criteria. However there is an added major benefit. Even good projects that do not go through a similar process can sometimes be slowed because there is not buy-in throughout the organization. R&D might champion a good project and work with marketing to continue development and then at the end manufacturing may balk and throw up roadblocks preventing commercialization. The Stage-Gate™ process ensures that the different functions are involved with the decision-making throughout. This helps immensely with final buy-in across the organization.

6.4 ORGANIZATION

In smaller companies, the way resources are organized is less critical but larger companies need to organize to maximize efficiencies. Two basic options are to organize by business or by function. When organized by business, everyone in the business group has a common direction and focus toward the business. The terms vary from company to company and even within a company based upon size of the business; it could be called a business group, business department, business division, etc. However, the business headsets the direction and everyone in the organization is held to that direction. This has the advantage in that there is a clear focus for individuals working within the business. Because they all work for the same business head, there is good cooperation between R&D, manufacturing, marketing and sales. Although organization by business may be best for that particular business, it may not be best for the overall company. Different businesses within a company do not have a strong motivation to help each other. In extreme situations, it can be competitive to the point of one business trying to take sales from another business even when they are part of the same company.

Therefore, some companies organize by function with all sales people reporting to the same sales manager, all manufacturing personnel reporting to the manufacturing manager, and so forth. Again the titles vary from company to company and the head of sales may be a vice president of sales, a sales director, a sales manager, etc. but the entire sales organization will report to this person. This organization by function has the advantage that a unified approach is presented by the sales force to external customers. Similarly, manufacturing may work together to maximize capacity utilization and R&D may work together to better share knowledge and talents across a wider variety of projects. The pitfall is that groups can lose focus on the overall business needs. The manufacturing group is aligned with each other and the goals of the manufacturing manager but these may not be aligned with the marketing and sales goals. For any type of organization, the most effective individuals are those that build relationships with others throughout their own and other functional areas.

There are some other organizational considerations. For example, the quality control (QC) group is responsible for testing materials to judge whether they are suitable for shipment to the customer. If a lot is defective, they need to have the authority to reject it. This can be contrary to the desires of a manufacturing group that is judged by pounds of product shipped. Therefore, it is best if the QC group does not report to manufacturing. Everyone in an organization has responsibility for quality but the quality assurance (QA) group has special overarching responsibility. They preach quality throughout the organization and are responsible for training and implementation of quality programs. QA is responsible to make sure that the tests run by QC are appropriate, done correctly, and relevant to customer needs. Because of their role, it is usually best if the QA group is given some type of autonomy and reports to upper management within the organization. Pharmaceuticals need to be manufactured according to cGMP, current Good Manufacturing Practices. These dictate such an organizational hierarchy for QA.

Within a group it is important that the necessary skills be present to accomplish the required tasks. Often this dictates that within a group there is a variety of people with different educational and experiential backgrounds. Just as you would not want to field a major league baseball team with all pitchers or all left-handed hitters, you want different skills present within the work group. Similarly, just as a team perfectly suited for baseball is probably terrible in hockey, a great basic research group might be terrible in quality control. A group of people with the correct skill sets, aptitudes and excitement for quality control might be terrible in basic research. It is common for a work group to have members with remarkably different training such as a collection of an organic chemist, a materials scientist, a chemical engineer and several experienced technicians, some of whom may have manufacturing experience. Education can range from Ph.D. to high school. Often there is an even greater diversity in personalities. When in such a group it is important for everyone to appreciate what the others bring to solve the tasks at hand. It is even more important to be able to communicate well with each other. Often in industry it is not necessarily the person that came from the most prestigious school or that had the best scholastic record that is the most effective. As simple as it sounds, the ability to work well with others and learn from others is a characteristic of the most effective employee.

Smaller groups of people can be organized in traditional hierarchical fashion with a group leader in charge. The title varies in different organizations and this role is often called supervisor or manager. The group leader is usually responsible for communicating to other groups and functions within the organization. Consider the example of a group charged with developing a process which can be scaled to meet a need identified by marketing. The group leader would be responsible to talk with marketing to make sure that all of the customer requirements are known and factored into the effort within the group. Similarly, the group lead needs to keep marketing and manufacturing appraised of progress and timetables so they can factor them into their plans. Often the group leader is senior and acts as a mentor for the group members. In some organizations, the group leader also gives appraisals, pay increases, disciplinary actions, and other tasks in concert with the human resources group.

Another option is to organize groups of people as self-directed teams. The advantage of this is that the team can assign tasks based upon skill set and time availability of the various team members. Even appraisals and salary actions can be done by the team based upon team ratings. For example a member is rated by everyone else on the team on a scale of several characteristics deemed to be important to the success of the team. Everyone could be rated on: technical expertise; work ethic; action oriented; cooperation; contribution to team goals; and other criteria. This team-based organization, or a variant of it, is commonly used. In theory, it is more effective because responsibilities are divided by the team and the team should best understand the capabilities of the members. However, it is not an ideal world and often there are issues with the non-hierarchical approach.

Whatever the organization, if trying to solve a problem, especially a plant problem, it is best to communicate directly with the people closest to the problem. These people often have the best understanding of the issues and if you take the time to listen to them will help you solve many a production problem. Similarly, if involved in scaling up processes, it is best to consult with the people in production at an early point in the development process to gain an understanding about what makes a successful process.

6.5 GANTT CHARTS

Many industrial projects involve people from different departments with different skills performing different tasks, each contributing to an overall goal. Management of these projects can be complex and often timing is important. Gantt charts can be used to plan and coordinate various needed tasks. They also offer a quick visual method of assessing whether a project is on schedule and what needs to be done to shift resources. Time is displayed on the X axis with various tasks on the Y axis. The tasks are represented horizontally to indicate the timing and needed time span. There are other techniques used in project management. For example, PERT (Program Evaluation Review Technique) charts are sometimes used because they can give a better sense of the interdependencies of the various tasks. However, Gantt charts are easy to understand and at a glance everyone can quickly understand how a project is progressing versus the schedule. They tell when tasks begin and end, how long they are expected to take, and what tasks overlap. They also give this in the context of the overall project.

Imagine a project that needs to move from early conception through laboratory work, scale-up and ultimately shipping product to the customer. This involves several functions within the organization and it is paramount that they each understand their role and how the project is progressing. The required tasks can be listed with some estimate of time needed to complete each one. This is represented in the table (Figure 6.1) and then as a Gantt chart (Figure 6.2).

It is easy for the pilot plant to look at the Gantt chart and realize that if the route optimization is delayed then they may have a late start. They either need to make up the time by perhaps doing their work in 14 days rather than the planned 28 or risk delaying the order for needed equipment for the manufacturing scale up. However, if they can give an early answer to the decision as to what equipment is needed, the other pilot plant work can be done somewhat later without impacting the overall timeline.

6.6 COST ESTIMATES

Projects need to be profitable. Profit is revenue less costs. The judgment of whether to proceed with a project should involve some consideration of profitability. Revenue projections can come from market studies and estimates of how much can be sold at what price. The other piece of the

Task	Start date	Duration	End date
Patent and literature review	1/7/2013	7	1/14/2013
Laboratory experiments	1/14/2013	21	2/4/2013
Finalize synthetic route	2/4/2013	2	2/6/2013
Optimize route	2/6/2013	21	2/27/2013
Pilot plant work	2/20/2013	28	3/20/2013
Order manufacturing equipment	3/20/2013	3	3/23/2013
Manufacturing scale-up	4/20/2013	14	5/4/2013
Ship to customer	6/1/2013	4	6/5/2013

Figure 6.1 Timetable of Required Tasks.

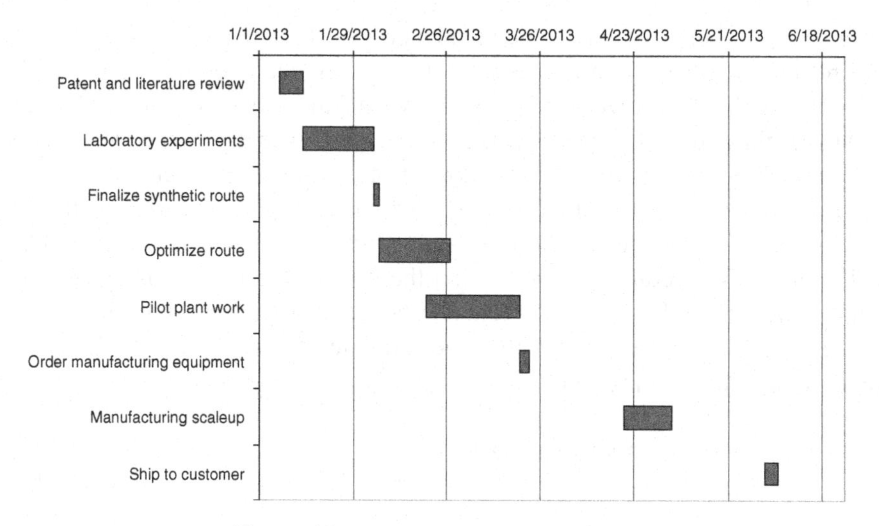

Figure 6.2 Gantt Chart of Required Tasks.

equation is cost. The cost includes many components, and detailed cost estimates are usually performed by people with extensive experience and training. The cost includes labor costs, both during the development and later during the production. Overhead is also included. Raw material cost is a major component. Because the amount of raw material cost varies with the amount of product produced, this is often considered as a part of variable cost. Another component is capital cost. This is money required to buy anything from an entire plant to necessary equipment. Because this cost is fixed and not dependent upon how much material you later make, this is sometimes called a fixed cost to contrast it from variable costs.

When thinking about capital costs, it is important to think about a concept sometimes referred to as the present value of money or the opportunity cost of money. This fluctuates based upon inflation and interest rates. Let's think about it with a very simple example. Imagine that if you spent $100, you could build a plant that could generate $120 beyond cost in ten years. You might consider a profit of $20 to be attractive. This is a 20% return on your initial $100. However, you need to remember that because you spent the $100, it was not available for other investments. Had you been able to invest the money in a bond paying 7.2% interest, at the end of ten years, you would have $200. So by investing in a plant and making $20, you have lost the opportunity to make $100 profit in a bond. You have effectively lost $80 in opportunity cost. This example is oversimplified and needs to reflect current interest rates as well as risks both with alternative investments and with investments for the project, but serves to illustrate that a dollar in the present is more valuable than a dollar in the future. This needs to be factored into a detailed cost estimate.

Other items such as interest, taxes and depreciation are also included in the detailed cost estimate.

The ratio of profit to investment is called return on investment (ROI). If there is a single up-front investment needed, it would be the return on original investment. A situation where continual investments are needed, such as for continual equipment upgrades, would be better calculated as return on average investment. Another common term used to determine the desirability of a project is the payout time. This is done by calculating the sum of profit plus depreciation each year until the sum equals the value of the investment. The lower the payout time, the more financially attractive is the project.

Because the raw material cost is a major component of a cost estimate, it is worthwhile to learn how to calculate this. If the raw material cost is higher than the anticipated selling price, there is a problem. If it is significantly lower than the selling price, the project probably warrants further consideration. Raw material costs can be done by simple calculations, sometimes referred to as "back of the envelope" calculations because they can be done on the back of an envelope during a discussion.

Imagine that we have an opportunity to sell ethyl acetate for $0.70 per pound and we know how to make it from acetic acid and ethanol. If we can achieve 100% yield and acetic acid costs $0.40 per pound and ethanol costs $0.38 per pound, is it worth considering the opportunity?

$$\underset{\text{Acetic acid}}{CH_3CO_2H} + \underset{\text{Ethanol}}{CH_3CH_2OH} \xrightarrow{H^+ \text{ catalyst}} \underset{\text{Ethyl acetate}}{CH_3CO_2CH_2CH_3}$$

You might be surprised at how many people will quickly say, "Of course not; it will cost you $0.78 and you can only sell it for $0.70." This would be true if one pound of acetic acid and one pound of ethanol make one pound of ethyl acetate. But that is not the case. One mole of acetic acid and one mole of ethanol make one mole of ethyl acetate.

$$\underset{\text{Acetic acid}}{CH_3CO_2H} + \underset{\text{Ethanol}}{CH_3CH_2OH} \xrightarrow{H^+ \text{ catalyst}} \underset{\text{Ethyl acetate}}{CH_3CO_2CH_2CH_3}$$

	Acetic acid	Ethanol	Ethyl acetate
Molecular mass	60 amu	46 amu	88 amu

By calculating molecular masses, we can say that 60 g of acetic acid can be combined with 46 g of ethanol to make 88 g ethyl acetate. This mass ratio holds true for different mass units. So 60 pounds of acetic acid can be combined with 46 pounds of ethanol to make 88 pounds of ethyl acetate. Similarly 60 tons of acetic acid can be combined with 46 tons of ethanol to make 88 tons of ethyl acetate. What about the conservation of mass? How is it that 60 + 46 = 106 pounds of starting material only makes 88 pounds of ethyl acetate? The answer is that we also make 1 equivalent of water; in this case, it is 18 pounds, so 106 pounds of starting material makes 106 pounds of product.

If we combine 60 pounds of acetic acid at a cost of $0.40/pound with 46 pounds of ethanol at a cost of $0.38/pound we spend $41.48 to make 88 pounds of ethyl acetate. This calculates ($41.48/88 pounds) to be a cost of about $0.47 per pound. Compared with an anticipated selling price of $0.70, that leaves $0.23 per pound for other costs and profit. This may not sound like much, but if it is a 10 million pound plant, it is $2.3 million. It is probably worth considering the project.

If the yield is 95%, we can divide the raw material costs by 0.95. So $0.47/0.95 = $0.49 per pound raw material cost. This is a less desirable situation because the lower the yield, the higher the raw material costs and also the larger amount of waste by-product that needs to be considered, often as purification or waste disposal cost.

As a second example, we can calculate the raw material cost of phenol. To keep the example simple, we will assume 100% yield and no value from the production of byproducts. For the example, assume that cumene costs $0.50 per pound.

Cumene $\xrightarrow{O_2}$ $\xrightarrow{H_3O^+}$ Phenol + Acetone

From the equation, we see that one mole of cumene makes one mole of phenol. Taking into account molecular weight, this means that 120 g of cumene (C_9H_{12}) makes 94 g of phenol (C_6H_6O). With the mass ratio of 120:94, we can say that it takes 120 pounds of cumene to make 94 pounds of phenol. Or that it takes 1.28 (120 / 94) pounds of cumene to make one pound of phenol. Therefore the raw material cost can be calculated: (1.28 lb cumene/ lb. phenol)($0.50 / lb. cumene) = $0.64 / lb. phenol. There are many other things which would need to be considered in a more thorough cost analysis. You also generate 0.62 (58 / 94) pounds of acetone per pound of phenol. The price you can get for the byproduct acetone will be a major factor in profitability of a phenol plant. You also need a radical initiator and oxygen in the first step. The amount of radical initiator used is very small and will not have any significance on the cost. Air is the source of oxygen and the cost is not significant. Acid and water are used in the second step. The acid is catalytic and therefore not of large cost significance. Water is relatively inexpensive.

6.7 SCALE-UP CONSIDERATIONS

Industrial chemicals are made on a large scale, often thousands of tons per day. Laboratory chemists typically work on gram scale, sometimes running reactions at the milligram level. The laboratory chemist usually uses glass flasks, often in very dilute solutions and with a lot of surface area. The laboratory reactions are usually run one at a time, also known as batch-wise. A reaction can perform well under these conditions but can have serious issues upon scale-up. There are many factors that go into a successful scale-up and the sooner the chemist thinks about them, the more effective they will be. My first boss [11] in industry was a chemical engineer and fond of telling the story of the lab chemist who accidentally broke a test tube containing a liquid on the bench and crystals formed in the puddle. The chemist then turned to the chemical engineer and stated, "My job is done; now you scale it up." The story is an overstatement and a way to jab at naïve chemists, but it is difficult not to be naïve as a recent graduate.

If you have several years' experience in scale-up, you will see a lot and will learn from what goes right and from mistakes. Chances are that many of you do not have decades of scale-up experience. If you are faced with scaling from the laboratory, my best advice is to seek counsel from those that are experienced.

Dependent upon the amount of material needed, the availability of equipment, and the confidence in the ability of the process to be run on a large scale, often an intermediate size operation is done. This can be a large-scale laboratory reaction, perhaps moving from 100 mL to a 22 L flask. Or it can be at a scale of several hundred gallons. Often these intermediate size processes are done in a facility expressly built for the purpose of trialing or piloting processes before eventual manufacturing scale. These facilities are called pilot plants. They often have more versatility in equipment than a manufacturing facility. However, sometimes they are built to exactly mimic the existing manufacturing plant.

One of my first scale-ups involved isolating a solid from slurry. In the lab, this was trivial; I just filtered with a Buchner funnel. I didn't care how long it took because I could do other things while it was filtering. In the plant and at a scale of thousands of gallons, it was important to be able to do an efficient separation. The available equipment was a centrifuge. Think about a large washing machine where after the wash the water and clothes are spun. The water passes though the washing machine basket, leaving the clothes behind.

The centrifuge to be used operated on a similar principle. It was lined with a cloth and spun. The slurry was then added to the spinning centrifuge. The liquid passed though the cloth leaving the spun solids behind. A blade would then scrape the solids off the walls of the centrifuge and they would exit the centrifuge through a chute at the bottom. One of my tasks was to identify a suitable cloth. Conceivably if the weave was too fine, the separation would be too slow. If too coarse, then fine particulate solids could pass through.

Coincidentally, another member of our group, also a freshly minted Ph.D. chemist, had the same task for a different process. We took different approaches. I took a bottle of laboratory-prepared slurry and brought it to the plant. I talked to the operators who ran the centrifuge and solicited their advice. I also consulted with some of the experienced people in our group. Based upon the advice given, I selected a cloth and trialed it on a laboratory centrifuge. It worked well and when done in the plant, the separation went smoothly.

The other chemist took a different approach. He did statistical laboratory experiments with dozens of cloth materials. After weeks of research and some very impressive weekly reports, he identified an optimum cloth. When the process went to the plant, the separation was a disaster. The material would not separate in the centrifuge. After several delays and switching cloths, the separation was finally done. The delays put the plant behind schedule and did not do much for the credibility of our group. All of this could probably have been avoided by consulting experienced people prior to the scale-up.

On the topic of communication regarding scale-ups, let me relate another true story. A scale-up I did early in my career involved the bromination of a ketone. It was done in a pilot plant on a scale of 100 gallons. The reaction proceeds through the enol so is acid catalyzed.

$$RC(=O)CH_2R' \underset{}{\overset{H^+}{\rightleftharpoons}} RC(OH)=CHR' \xrightarrow{Br_2} RC(=O)CH(Br)R' + HBr$$

A solution of the ketone in water is made. A small amount of HBr is added and then bromine gradually added. The reaction is rapid and exothermic but can be controlled by the rate of bromine addition. For each equivalent of bromine added, an equivalent of HBr is formed. Initially the HBr dissolves in the water but eventually the water becomes saturated and the HBr evolves as a gas from the reaction. In the lab, it is easy to contain the HBr. I passed the gas through a flask containing a sodium hydroxide solution. In the pilot plant, I planned to pass the gas counter-current through a water stream. The water absorbs the HBr gas. Based upon HBr solubility in water and bromine addition rates, I was able to calculate the water flow needed. This process of absorbing an effluent gas is called "scrubbing." A gas can be scrubbed with either a liquid or solid.

Laboratory scale reactions are often done during a normal work day. On a larger scale, things can take longer and equipment is often in demand. Therefore larger scale reactions are typically run in shifts and around the clock. Everyone understood the plan for the bromination and we began working in shifts. When I arrived for my shift, acidic gas was exiting out the roof vent of the pilot plant. When I got inside, I was greeted with, "the reaction was going so well, we decided to up the rate of bromine addition." They had done this without also upping the water flow in the scrubber. Therefore the water was not able to absorb the HBr and was venting out of the roof. We immediately stopped the bromine addition and got things under control.

There are two lessons to be learned from this mistake. One is that although it is obvious to a chemist that the rate of HBr evolution is tied to the rate of bromine addition, it is not necessarily obvious to everyone (in this example an experienced chemical operator and a chemical engineer). I should have done a better job in communicating this ahead of time. The second lesson is that procedures should not be arbitrarily changed during a scale-up. This problem was quickly remedied before any significant harm was done but it could have been much more serious. There should be strict guidelines about any process changes and what approvals are needed before the changes are made.

There are many things to consider when scaling a process and many distinctions between large-scale chemical reactions and small laboratory scale reactions. I will discuss some of them here. The discussion is not meant to be exhaustive but should serve to get the reader to begin thinking about some of the differences. Each reaction will have its own extra things to be considered. Some extra consideration should be given to raw materials, chemicals used, heat exchange, safety, material of construction, waste, agitation, and time. Often, there is a need to use existing equipment so the process needs to fit the equipment and not vice-versa.

Raw materials are often overlooked but sometimes can cause unexpected results. Often in the laboratory, a single source or even a single bottle of a starting material is used. When scaled, larger quantities are used. The starting materials are often purchased from different suppliers and not laboratory supply vendors. The quality can vary from what was used in the laboratory. Sometimes, even minor differences in quality can have an important and unexpected effect on the process. I was involved with one product that had been successfully made and sold in Europe. When the manufacturing needs were so large that it also needed to be made in the U.S., there were terrible quality issues. After months of fighting the problem, we learned the quality of one of the starting materials was slightly different. Surprisingly, the purer product used in the U.S. gave poorer results because a minor impurity in that ingredient was critical for success.

Many chemicals are relatively easy to use in the laboratory but because of reactivity, flammability, or toxicity can present problems on scale-up. They can require special procedures or equipment and can increase the cost of the process. It is sometimes appropriate to completely change a synthetic sequence in order to avoid the use of certain chemicals.

Laboratory reactions are often done in a glass flask with plenty of surface area. The reaction can be heated or cooled by placing the flask in a container containing a hot or cold liquid. To cool, ice baths are commonly used. Baths containing an organic solvent and dry ice such as dry ice/acetone baths are also routinely used if further cooling is warranted. Flasks can be heated with hot water, steam, or by immersing them in hot oil. There is a multitude of options in the laboratory.

There are fewer practical heat-transfer options on scale-up, especially if required to use existing equipment. Reactions are often done in large volume reactors which have much less surface area per reactor volume than the laboratory flask. Because of this, heat transfer is often much less efficient. The heating or cooling is often done by use of a jacketed reactor. A liquid is pumped through a jacket which surrounds the reactor and provides heating or cooling. Often there are limitations as to the heat transfer fluid to be used. For example, a reactor equipped with cooling water is not typically also used with hot oil circulating in the jacket.

It is a lot easier and faster to cool from 70 °C to 10 °C if you have 50 mL in the laboratory setting than if you have 3,000 gallons in a plant reactor. The time factor can be important if there are stability issues with your reaction mixture. It is important to recognize that operations can take much longer and to plan for this. One scale-up I experienced involved a chemical that would degrade in the liquid state. In the laboratory, it was purified, cooled to solidify and stored in a refrigerator. When needed, it was melted just prior to use. This was trivial in the laboratory but disastrous upon scale-up. After larger scale purification, it took many hours to cool it to solidify and there was some loss of purity. Then after storage, melting was a very long process and the purity suffered more. The lesson is that heat transfer limitations need to be given serious consideration prior to scale-up.

Heat transfer is even more serious for an exothermic reaction. Minor exotherms in the laboratory may not even be apparent or if they are noticed can be easily controlled with cooling. However, the same exotherm on a large scale can be difficult to control and even present a safety hazard. Therefore it is important to thoroughly understand the thermodynamics before increasing reaction scale. This is often done by calorimetry experiments.

In addition to heat transfer, there are many other scale-up considerations from a safety perspective. Flammable solvents and gases present a risk because of the possibility of sparks in a plant environment. When possible, they should be avoided. If not possible, precautions need to be taken. Every effort needs to be taken to contain the chemicals. This is especially important for the safety of workers but also to protect the environment. Serious consideration needs to be given to generated waste. It should be minimized but at the very least there needs to be a responsible plan on how to treat it.

Material of construction is a topic that chemists may not be used to thinking about. In the laboratory, reactions are run in glassware and there are rarely issues. Plant equipment is often carbon steel or stainless steel. If your reaction generates or uses a corrosive material such as aqueous HCl, then there are major issues and this equipment should not be used. For reactions where steel is not suitable some alloys such as Monel or Hastelloy may be suitable. Sometimes nickel is used. More commonly, a steel reactor is lined with glass, effectively making this a glass reactor. Sometimes the glass lining can have chips or holes due to thermal shock, abrasion or even dropping tools in a reactor. The glass lining can be tested for thickness by using a magnetic induction instrument. Defects can be found by conducting a spark test. In special circumstances, reactors are lined with Teflon.

Sometimes the material of construction is a factor in the reaction. Consider the free radical chlorination of toluene to make benzyl chloride.

$$\text{Toluene } (C_6H_5CH_3) + Cl_2 \xrightarrow{\text{Light}} \text{Benzyl chloride } (C_6H_5CH_2\text{-}Cl) + HCl$$

The reaction should not be done in a steel reactor because of corrosion issues. But there is another reason. Even the slightest bit of corrosion in the presence of chlorine will give iron (III) chloride. This serves as a Lewis acid and catalyzes electrophilic aromatic substitution. The reaction is drawn showing the formation of the para isomer but other isomers are also formed. Ring-chlorinated toluene is an unwanted and difficult-to-remove impurity if you are making benzyl chloride.

$$\text{Toluene } (C_6H_5CH_3) + Cl_2 \xrightarrow{FeCl_3} p\text{-}ClC_6H_4CH_3 + HCl$$

Especially in situations where trace metals can change the chemistry, all of the equipment needs to be considered. It is not enough to use a glass-lined reactor if the piping feeding the reactor is made from carbon-steel.

Ben Franklin is credited with the saying, "Waste not; want not." I don't think that he was referring to chemical waste but chemical waste is a major factor in scale-up. In the laboratory, waste from filtrates, unused distillation fractions, etcetera is of inconsequential quantity but as the scale increases, it can mean the success or failure of a project. In the laboratory, reactions are often run in dilute solutions. On a larger scale, to minimize waste and to be able to make more material, reactions are commonly much more concentrated; sometimes they are run without solvent. When a solvent is used, it is best if it can be recycled. If the solvent must be disposed of, it adds to the cost and is not as environmentally friendly. When there is less solvent to act as a heat sink, heats of reaction become more pronounced. Lacking other controls, an exothermic reaction run at 20% concentration will have a much greater temperature rise than one run at 2%. The characterization of the waste is also important. Whenever possible the waste should not be contaminated with highly toxic or environmentally unfriendly chemicals.

If a reaction success is highly dependent upon agitation, it can present a challenge in scale-up. Examples of reactions like this might include those occurring at an organic / aqueous interface. Often agitation in the laboratory is more effective than in a large-scale continuously stirred reactor (CSTR). When extra agitation is needed, it can be accomplished by reactor design, but the laboratory chemist must realize that it is important.

Time is something else that can vary on scale-up. It is easy to rapidly heat or cool reactions. An ingredient can be added to the reaction in a second. Typically, things take longer on a larger scale. It is important to realize this and perhaps test the importance in the laboratory so that there are no surprises on scale-up.

Another major decision on scale-up is batch versus continuous. If something is to be manufactured in a batch reaction, starting materials are added, the reaction is performed, and the product purified and isolated. If more is needed, then a second run or a second batch is performed. Often batch manufacturing processes run one batch after another around the clock. Another option is to run chemical reactions on a continuous basis. In a continuous process, starting materials are continuously fed to the reactor and product continuously purified and isolated.

We can consider the two options by thinking about the production of ethyl acetate. In a batch production acetic acid, ethanol, and catalyst are added to a reactor, the mixture is heated for a period of time and then distilled. The different distillation fractions are collected. This process is repeated as often as necessary to meet customer demand. Each batch produced has its own identity. In a continuous process acetic acid, ethanol, and catalyst are continuously fed to a heated reactor. At the same rate, crude reaction mixture is continuously removed from the reactor and continuously fed to a distillation column. At the same rate, the fractions are continuously removed from different sections of the column, with more volatile materials continuously exiting the top of the column and less volatile chemicals continuously exiting the bottom of the column. Ideally, the streams are also analyzed on a continuous basis.

$$\underset{\text{Acetic acid}}{CH_3CO_2H} + \underset{\text{Ethanol}}{CH_3CH_2OH} \xrightarrow{H^+ \text{ catalyst}} \underset{\text{Ethyl acetate}}{CH_3CO_2CH_2CH_3}$$

Each method has its advantages. It is easier to start up and shut down a batch process. Batch processes are desirable if many different materials are to be made in the same equipment. If the reaction causes a coating or scale on the reactor walls, a batch process is more desirable because reactors can be cleaned between batches. Batch processes are also best if the reaction is not performed on a 24 hours per day and seven days per week schedule. If something goes wrong in a batch reaction, you can stop and investigate the problem prior to starting the next batch.

If the same or a very similar material is to be made in the same equipment for an extended period of time, often a continuous process is better. There are not the long lags of time lost as a batch process is begun or ended. Once the process is on stream, product is continuously being made. There is a better opportunity for consistent quality. Once the parameters of the process are set, the reaction can run for extended periods of time making the same very uniform material. Typically, more material can be made in the same size equipment when a continuous versus batch process is used. Usually, less manpower is needed to keep a continuous process running versus the demands of starting and finishing batch-wise reactions. In one example, a drug firm, Dr. Reddy's, replaced a batch process with a continuous process and achieved an annual savings of $1.1 million [12].

Even if property differences are needed for the marketplace, they can often be accommodated in a continuous process by making adjustments to the operating parameters without stopping the process. For example, polyethylene is made by a continuous process yet the same plant will make different grades that vary in density. This is done by adjusting the input ratio of comonomer, such as 1-hexene, to monomer ethylene. There will be a time during the transition when intermediate density polyethylene is made. With good process controls and gradual changes in targeted density, these transition times and the amount of off-specification material can be minimized.

QUESTIONS

1. Assuming adipic acid costs 50.0¢ per pound and methanol costs 20.0¢ per pound, and assuming 100% yield, calculate the raw material cost of dimethyl adipate to the nearest tenth of a cent.
   Unbalanced equation:

$$HO_2C(CH_2)_4CO_2H + CH_3OH \longrightarrow H_3CO_2C(CH_2)_4CO_2CH_3 + H_2O$$

2. If methanol costs 35.0¢ per pound and terephthalic acid costs 45.0¢ per pound, and assuming 100% yield, what is the raw material cost of dimethylterephthalate (report your answer to three significant figures)?

3. Assume that the amount of base and N-acetyl caprolactam used is so small that it is insignificant in the weight of the polymer and in the cost. Nylon 6 is a polyamide made by the ring opening of caprolactam ($C_6H_{11}NO$) incorporating each of the atoms of caprolactam into the final polymer. Assuming 100% yield, if the cost of caprolactam is 68.0¢ per pound, what is the raw material cost of nylon 6 (report your answer to three significant figures)?

4. Choose the best answer. Although everyone in an organization has a responsibility to deliver a quality product, this group preaches quality throughout the organization and is responsible for training other groups in quality.

   a. R&D
   b. QC
   c. QA
   d. Manufacturing
   e. QRS

5. You and Edith work in the R&D group. The plant manager tells Edith that they are having a problem with variability in the granulation of valproic acid (Depakote). Production uses a melt granulation method with binding agents. Edith's specialty is mass spec so she is thinking of running mass spec on the binding agents. Edith often comes to you for advice. What would you tell Edith? (Answer in 20 words or less).

REFERENCES

1. Greg Brue, Robert Launsby. *Design for Six Sigma*. McGraw-Hill, United States; 2003:3.
2. Greg Brue. *Six Sigma for Managers*. McGraw-Hill, United States; 2002:29.
3. Greg Brue. *Six Sigma for Managers*. McGraw-Hill, United States; 2002:29.
4. Thomas Pyzdek. *The Six Sigma Handbook*. McGraw-Hill, 2nd edition, United States; 2003:238.
5. Thomas Pyzdek. *The Six Sigma Handbook*. McGraw-Hill, 2nd edition, United States; 2003:28–29.
6. BC Cloete, A Bester. Onderstepoort Journal of Veterinary Research 2012; 79(1), Available at: http://www.ojvr.org/index.php/ojvr/article/view/407. Date accessed: 29 Apr. 2013.
7. Alex Dmitrienko, Christy Chuang-Stein, Ralph B D'Agostino. *Pharmaceutical Statistics using SAS: a Practical Guide*. SAS Publishing, Cary, NC; 2007:4.
8. John LaMattina. *Drug Truths*. New York: Wiley and Sons Inc., Hoboken, NJ; 2009:16–17.
9. R Cooper, S Edgett, E Kleinschmidt. *Portfolio Management for New Products*. Addison-Wesley, Reading, MA; 1998:4.
10. R. Cooper. *Winning at New Products*. 4th ed. Basic Books, New York, NY; 2011.
11. Dr. Patrick McDonald, personal communication, 1978.
12. Alex Scott. Chemical and Engineering News 2013; 91(21):30.

CHAPTER 7

Polymer Basics

"Mer" is derived from the Greek word "moroε" meaning part. Therefore, a monomer means one part or one unit; dimer means two units; oligomer means a few units; and polymer means many units. A polymerization involves the reaction of a monomer to connect many monomer units. If the reaction is additive, where one monomer adds to the next as in the polymerization of styrene, the polymer is called an addition polymer. Polymerizations of olefins, such as ethylene, propylene, vinyl chloride, or styrene, are addition polymerizations. In an addition polymer, all of the atoms of the monomer remain in the polymer. If the monomers are connected in a condensation reaction such as when a carboxylic acid and an alcohol react to remove water and form an ester linkage, then the polymer is called a condensation polymer. Recognize that in this type of polymer all of the monomer atoms are not incorporated into the final polymer.

One common method used to polymerize olefins is a free radical polymerization. For example, polystyrene can be prepared by heating styrene with a small amount of a radical initiator such as benzoyl peroxide. Benzoyl peroxide forms the benzoyl oxy radical and the phenyl radical [1]; for simplicity only the benzoyl oxy radical is shown.

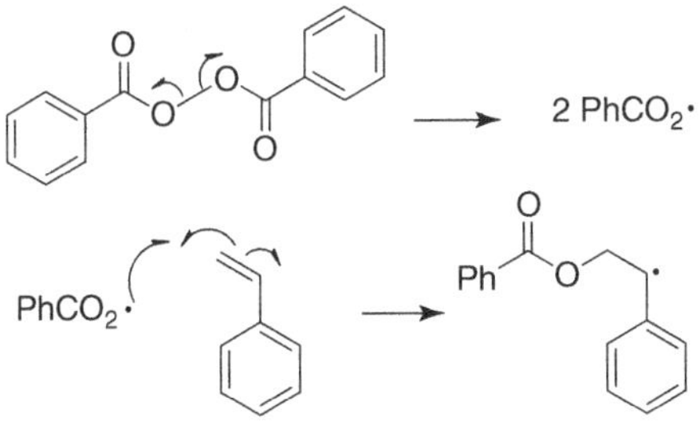

$\longrightarrow$ 2 $PhCO_2\cdot$

$PhCO_2\cdot$ $\longrightarrow$

Fundamentals of Industrial Chemistry: Pharmaceuticals, Polymers, and Business, First Edition. John A. Tyrell.

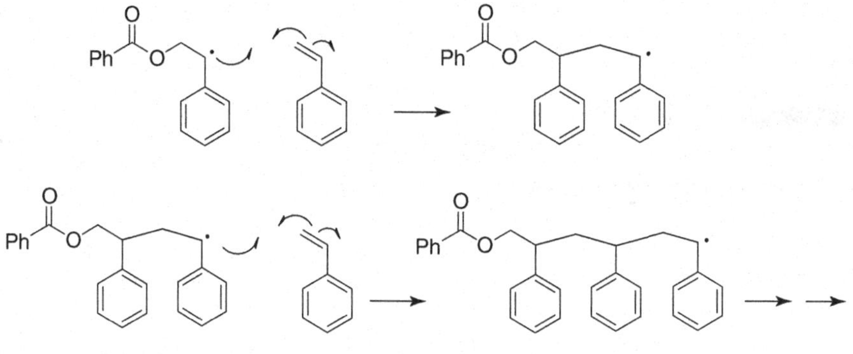

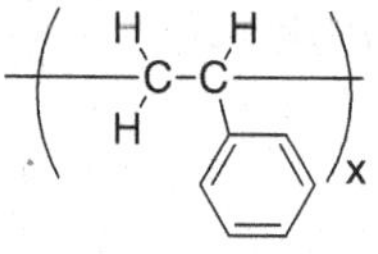

Polystyrene

The benzoyl radical reacts with styrene to form the benzylic radical. This reacts with a second styrene molecule to form the dimer. The reaction continues through a series of propagation steps to form ultimately polystyrene. Rather than writing the repeating structure over and over, it is written in parentheses with a letter such as m, n, x, y, or z to indicate an unknown number of repetitions. If this method is used to write the dimer, the number 2 would be used in place of x. Note that all 16 atoms of styrene (C_8H_8) are incorporated into the polymer.

Radical polymerization of styrene is the predominant method of polystyrene manufacture, but styrene can be polymerized by other means. Styrene can be polymerized by an anionic mechanism, for example by treatment with a minor amount of butyl lithium to initiate the reaction.

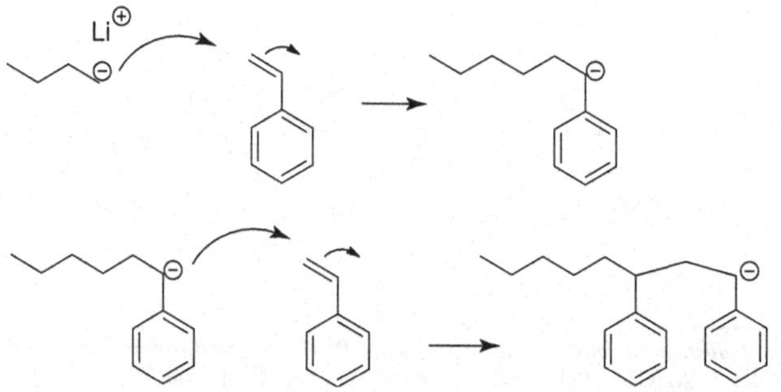

In the anionic mechanism, the carbanion reacts with styrene to form a new benzylic carbanion. This reacts with another molecule of styrene to form the dimer, which then continues to add styrene molecules to form, eventually, polystyrene.

Another polymerization technique is cationic polymerization. For example, styrene can be polymerized by treatment with a small amount of 1-phenethyl chloride, tin tetrachloride, and tetra-n-butylammonium chloride [2]. Presumably, the mechanism is as follows.

Polystyrene can also be made by using organometallic catalysts to polymerize styrene. Polymers can have stereochemistry and the use of certain organometallic catalysts is a means to control stereochemistry. Consider two conformers of 1,3-diphenylpropane.

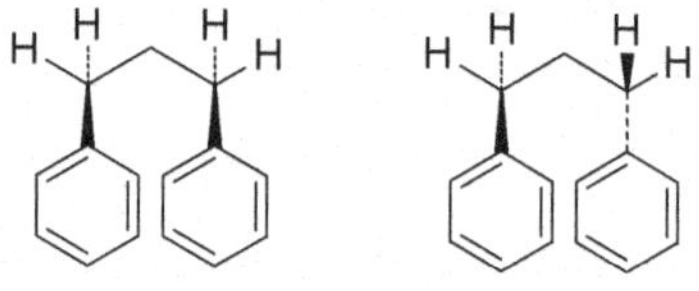

Because there is free rotation of the single bonds, these conformers rapidly interconvert and there is only one 1,3-diphenylpropane. These two conformers are not different compounds. Just as a person walking down the street remains the same person and we don't have one person with the left foot forward and another person with the right foot forward, the conformers are different representations of the same molecule.

Now consider 2,4-diphenylhexane. The two benzylic carbons (the carbons directly attached to the phenyl groups) each have four different groups attached to them, and therefore they are each chiral carbons. There is no plane of symmetry in the molecule and therefore there are different stereoisomers. The two stereoisomers depicted below (they are diastereomers) cannot be interconverted by bond rotation.

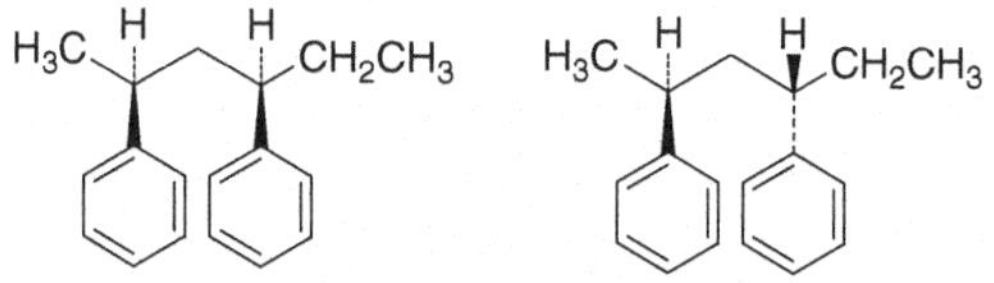

In the case of a polymer, we can have a similar situation. Consider the benzylic carbons on two neighboring units within a polystyrene polymer chain, arbitrarily designated as carbon "a" and carbon "b" in the structure. These carbons are asymmetric or chiral. In this structure, the phenyl groups on carbons a and b are on the same side of the polymer chain.

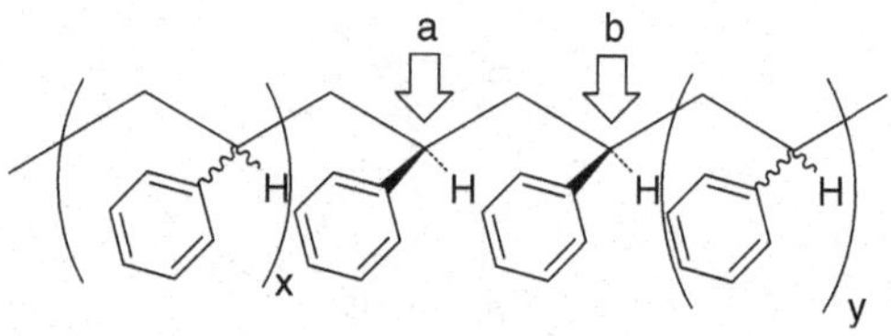

Now, let's consider the same polymer with the stereochemistry inverted at carbon b. The squiggly lines in the repeating structures mean that the direction is not assigned.

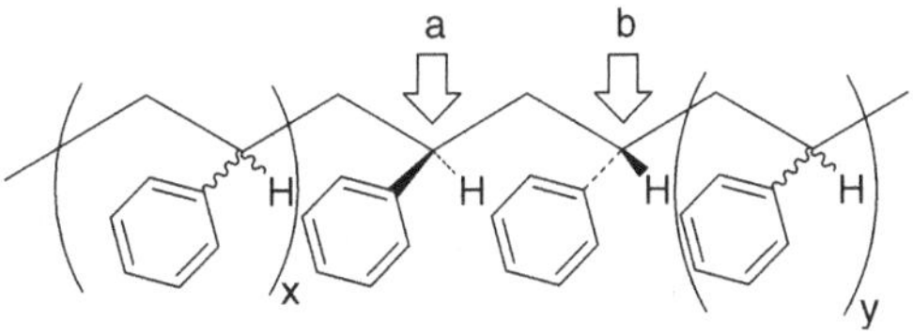

Now, the phenyl groups on carbons a and b are on the opposite side of the polymer chain. This is a different structure and one cannot be converted to the other without bond breaking. Because of this, polystyrene can have different structures depending on which side of the ring the phenyl group is located. Now consider the stereochemistry along an entire chain. When a polymer structure is written, typically only one unit is written with parentheses or brackets indicating that the unit repeats. However, to illustrate the stereochemistry, several units are depicted. In this structure, all of the groups attached along the polymer chain are on the same side of the polymer. The stereochemistry of a polymer is called the tacticity and a polymer with all groups on the same side is called isotactic.

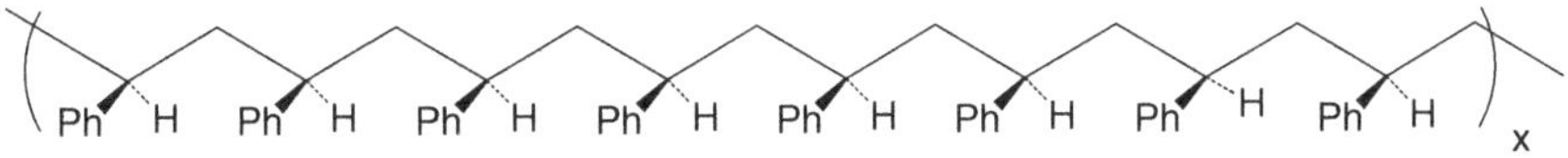

Isotactic polystyrene

If the groups are randomly arranged, the polymer is said to be atactic. When the groups alternate from side to side, the polymer is syndiotactic.

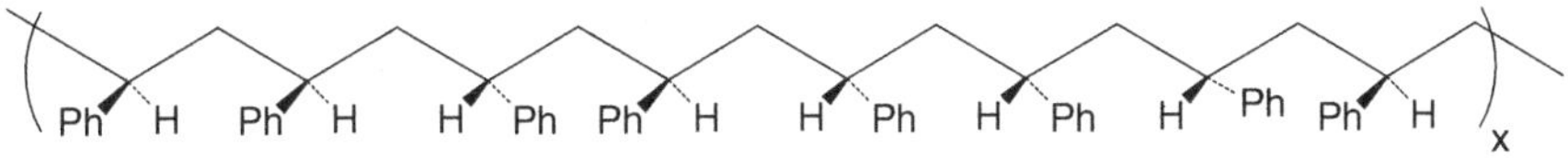

Atactic polystyrene

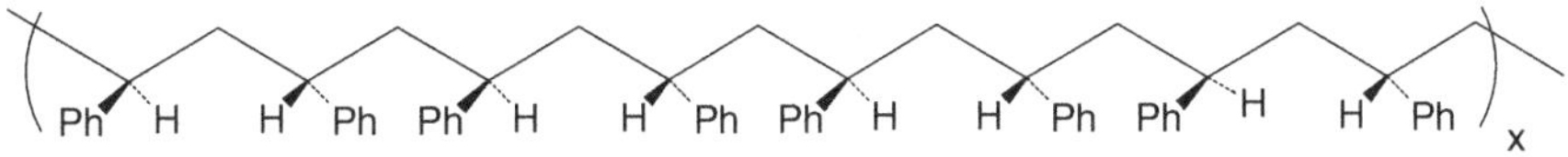

Syndiotactic polystyrene

The olefin polymers such as polystyrene, polyethylene, and polypropylene are all addition polymers. In an addition polymer, the monomer units are "added" together with each of the atoms of the monomer incorporated into the polymer. The other major type of polymer is condensation. In the formation of a condensation polymer a molecule, often water or an alcohol, is eliminated when each polymer bond is formed. For example, a polyester made by the reaction of a dicarboxylic acid with a dialcohol is a condensation polymer because a water molecule is removed when each ester bond is formed.

Polymers can be made with a single repeating unit; these are homopolymers. The repeating unit can be varied to form copolymers. For example, if ethylene is copolymerized with propylene, a copolymer can be formed. Depending on the reaction conditions and the reactivity of the different monomers, copolymers can be alternating, random, or block. This is illustrated with polymers made from monomers X and Y.

—X-X-X-X-X-X-X-X-X-X-X-X-X—
Homopolymer of X

—X-Y-X-Y-X-Y-X-Y-X-Y-X-Y-X—
Alternating copolymer

—X-Y-X-Y-Y-X-X-X-Y-X-Y-X-X—
Random copolymer

—X-X-X-X-X-X-X-Y-Y-Y-Y-Y-Y—
Block copolymer

Another type of copolymer is a graft copolymer with one polymer chain built or grafted onto an existing polymer. There are many variations of these themes and copolymers are not limited to two monomers.

```
                        Y
                        |
                        Y
                        |
                        Y
                        |
                        Y
                        |
                        Y
                        |
                        Y
                        |
—X-X-X-X-X-X-X-X-X-X-X-X-X—
        |
        Y
        |
        Y
        |
        Y
        |
        Y
        |
        Y
        |
        Y
```

Graft copolymer

Most commercial polystyrene is made by a radical process which yields atactic polystyrene. Cationic and anionic polymerizations also yield atactic polystyrene. However, the use of organometallic catalysts enables the formation of atactic, isotactic, or syndiotactic polystyrene depending on the organometallic catalyst.

The use of organometallic catalysts dates back to the 1950s and remains an active research area to this day. Prior to the 1950s, all polyethylene was made by a high pressure, high temperature radical process. Because of the high temperatures and the indiscriminant nature of radical reactions, the polyethylene prepared by this process is not perfectly linear, but rather has some branch points. This prevents the polyethylene chains from packing as tightly and the resultant polymer has lower density and crystallinity. It is called low density polyethylene (LDPE). In the early 1950s, Karl Ziegler found that if an aluminum alkyl compound (now often referred to as the cocatalyst) were added to a titanium or vanadium compound (the catalyst), ethylene could be polymerized at low pressures. Much research has been done and it remains an active research subject to this day. A search of U.S. patents reveals that since 1975 more than 22,000 U.S. patents were issued containing the word "Ziegler." Shortly after Ziegler's work, Guilio Natta, working independently, applied these catalysts to the polymerization of propylene and was able to make isotactic polypropylene. In 1963, Ziegler and Natta were awarded the Nobel Prize in chemistry. The catalyst systems are referred to as Ziegler–Natta catalysts and most polyethylene and polypropylene produced today is made with Ziegler–Natta catalysts. For a variety of reasons, polyethylene is the largest volume plastic in the world. Polypropylene and polyvinyl chloride have similar manufactured volumes and are second and third after polyethylene. Because of the large volumes of polyethylene and polypropylene produced, there has been and continues to be considerable development of these systems and many variants have been discovered. A typical Ziegler–Natta catalyst system might comprise a titanium tetrachloride catalyst and a triethyl aluminum cocatalyst. Often, the catalyst is supported, for example on magnesium chloride. Sometimes, donors such as esters are added to control the stereoregularity. Recognize that because there are no groups on the polyethylene homopolymer chain, tacticity of polyethylene is not a factor, but this is important for polypropylene manufacture.

Ziegler–Natta catalysts are very important, have been known for more than 60 years, and have been the subject of many studies. There are several different active catalytic sites. Because the different active sites have different reaction kinetics, there is a wider molecular weight distribution than found in many other types of polymerizations. However the details of the active catalytic sites are not precisely known. This can be because they are heterogeneous and therefore more difficult to study by spectroscopic means. In the case of an olefin polymerization using $TiCl_4$supported on $MgCl_2$ and treated with diethyl aluminum chloride cocatalyst, it is known that the active center is the product of reactions between the Ti species and the cocatalyst. The ethyl groups from the diethyl aluminum chloride form the initial starting ends of the polymer chains. The growth reaction is an insertion of the olefin into the Ti – C bond in the active center. The active centers are coordinatively unsaturated, and therefore easily poisoned by ligands such as CO, phosphines and amines. However, other aspects of the polymerization mechanism and structure of the active centers are less certain [3].

Another type of olefin polymerization catalyst system, often referred to as a Phillips type catalyst, is based upon chromium trioxide, typically impregnated on a solid support such as silica or alumina [4].

In the 1970's Kaminsky, Sinn, and others discovered that bis(cyclopentadienyl)dimethyltitanium when mixed with trimethyl aluminum and water provided a catalyst system capable of polymerizing ethylene [5]. The titanium structure bears some resemblance to that of ferrocene. Ferrocene was reported in 1951[6, 7] and the following year, the correct structure reported [8]. A brief, interesting account of the early days of the research on the structure proof of ferrocene has been written [9]. Because the Kaminsky catalysts have the same "sandwich" structure of ferrocene, they are referred to as metallocene catalysts. Just as the ferrocene ushered in a new era of organometallic chemistry, the Kaminsky metallocene spurred a tremendous amount of research in olefin polymerization catalysts.

Ferrocene Bis(cyclopentadienyl)dimethyltitanium

The research has varied the metal, the non-coordinating methyl ligands, and the coordinating anionic pi systems. The combination of trimethyl aluminum and water is considered to be an in-situ source of methyl alumoxane (MAO), which can both methylate the metallocene and form a metal cation. Common metallocene catalyst systems now use a zirconium dichloride and MAO. MAO is often represented by the formula $(CH_3AlO)_n$ but this is an oversimplification and it is likely that MAO has a cage-like structure [10] and that the ratio of methyl to aluminum is greater than 1.0.

The polymerization can be rationalized by the following sequence. The zirconium dihalide is methylated by MAO and the MAO forms a zirconium cation, with the negative charge being dispersed along the MAO framework. MAO serves as a non-reactive anion. The zirconium cation is a strong electrophile and coordinates with the olefin; ethylene in this example. In an insertion step, the methyl group is transferred to the olefin and a propyl zirconium cation is formed. Another ethylene complexes with the zirconium cation and after the transfer of the propyl group, a new zirconium cation is formed. This process repeats and after thousands of iterations polyethylene is formed.

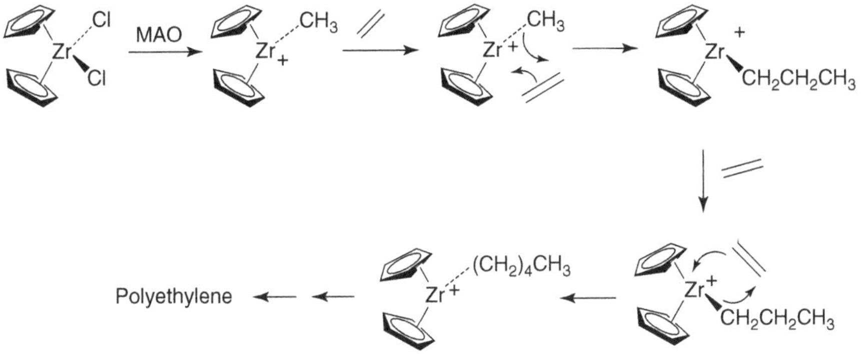

In the case of the homopolymerization of ethylene, there is no tacticity, but with propylene polymerizations the catalyst symmetry can influence the tacticity. Bis-cyclopentadienyl zirconocenes are symmetric and give atactic polypropylene. The isopropyl bridged cyclopentadienyl fluorenyl zirconocene has a center of asymmetry at the metal (C_s symmetry) and provides syndiotactic polypropylene. The same catalyst, but with a t-butyl substituent on the cyclopentadienyl ligand has C_2 symmetry and yields isotactic polypropylene [11].

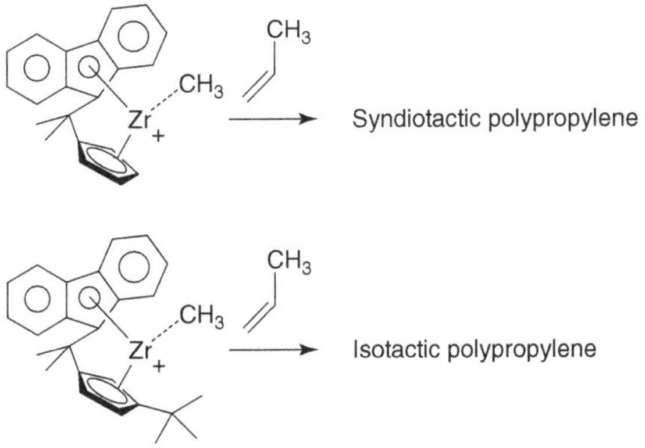

Because of the commercial significance of polyethylene and polypropylene and to a lesser extent, polystyrene, and the ability to tune polymer properties such as stereochemistry, molecular weight, and comonomer incorporation, research on these organometallic catalyst systems continues to this day. Metallocene catalyst systems are often based upon substituted cyclopentadienyl, indenyl, or fluorenyl systems. Often, as in the above example, the aromatic anionic ligands are bridged to each other.

Other systems may have one aromatic anionic ligand. These are not strictly metallocene catalysts, but are more properly referred to as single-site catalysts. They provide a single active site for the polymerization as opposed to Ziegler–Natta systems which have many active sites. Because there is a single active site, they tend to produce polyolefins with a narrower molecular weight distribution. The molecular weight distribution has a strong influence on certain properties such a viscosity versus shear and impact. One example of a single site organometallic catalyst is the system developed by Dow Chemical [12].

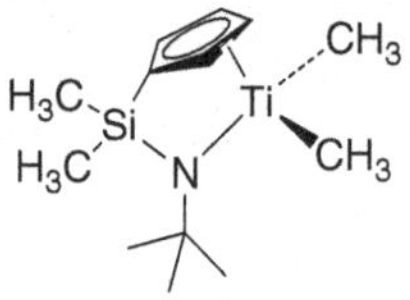

Dow constrained geometry catalyst

Equistar researchers have developed systems based upon heterocyclic pi-anionic ligands such as indenylindolyl [13] or boratabenzene derivatives [14].

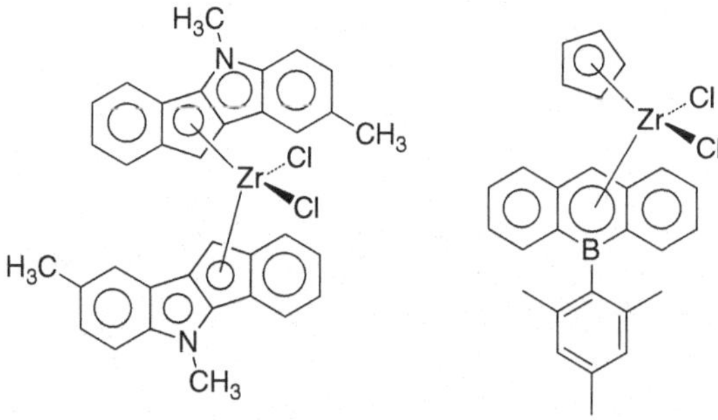

The choice of the catalyst can influence many physical properties such as toughness, processability, and heat capabilities. These properties are determined by the molecular architecture, things such as molecular weight, stereochemistry, comonomer incorporation, etcetera. The ability to change the catalyst to make a modified polymer suitable for new applications is of great commercial significance and ensures continued research in this area.

Another type of polymerization that also uses organometallic catalysts is ring-opening metathesis polymerization (ROMP) [15–17]. This is not used for the major volume polymers but is discussed here because it represents another method of polymerization. The word, "metathesis" derives from the Greek meaning to transpose or changing places. The olefin metathesis reaction is widely used in organic synthesis. Yves Chauvin, Robert Grubbs, and Richard Schrock were awarded the 2005 Nobel Prize in Chemistry for their work in this area. Simply viewed, olefin metathesis is an olefin exchange as exemplified by the general olefins a=b and c=d or by the reaction of 2,3-dimethyl-2-butene with 3,4-diethyl-3-hexene.

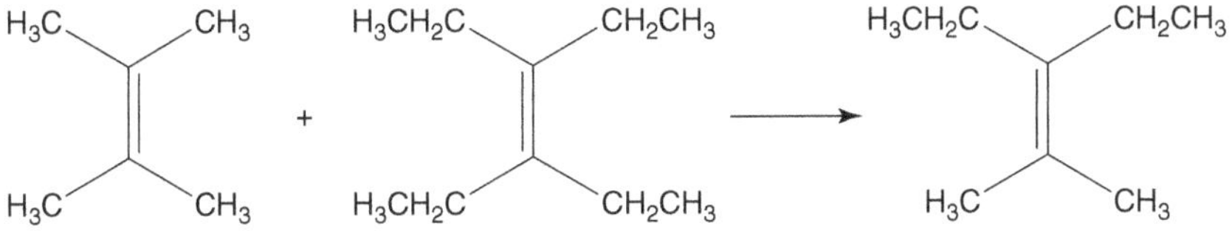

If the olefin is contained within a ring, preferably one with ring strain, it can react with itself by ring opening and the newly formed olefin is a polymer. This is exemplified by the ring-opening polymerization of norbornene.

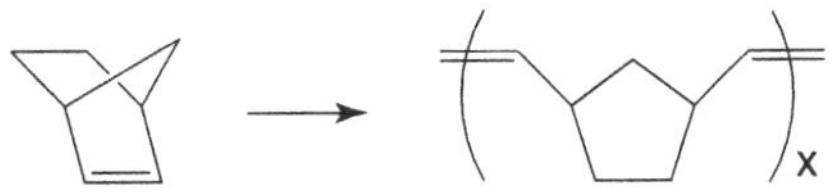

Note that although most olefin polymerizations "consume" the double bond, here the unsaturation is maintained. The polymer repeat unit in polyethylene or polystyrene does not include an olefin, but there is a repeat olefin in ROMP polymers.

Another polymerization type is living polymerizations. In a living polymerization, the polymerization proceeds to full conversion with further monomer addition leading to continued polymerization. This means that the growing polymer chain does not terminate, but remains reactive or alive. Polymerizations can terminate by chain transfer reactions or in the case of radical polymerization by combination or disproportionation.

Consider the radical polymerization of styrene. Two growing polymer chains can combine. This terminates the polymerization.

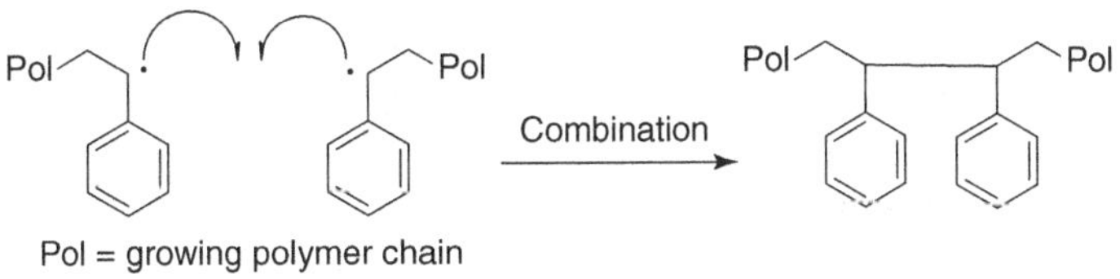

Pol = growing polymer chain

Another possible way for a growing chain to terminate is if the radical abstracts a hydrogen atom from another growing chain. This termination mechanism is called disproportionation.

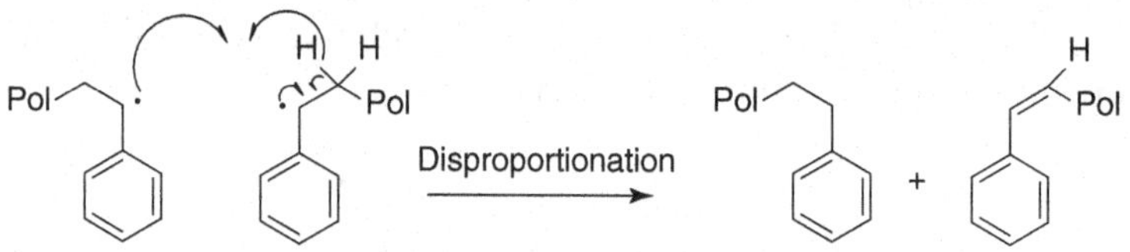

Pol H H Pol Disproportionation Pol + H Pol

Last, a chain can terminate by abstracting a hydrogen atom from a chain transfer agent. The chain transfer agent can be present as an unwanted impurity. Often a chain transfer agent is deliberately added to a polymerization to control molecular weight by terminating the reaction. The reaction is illustrated with butyl mercaptan.

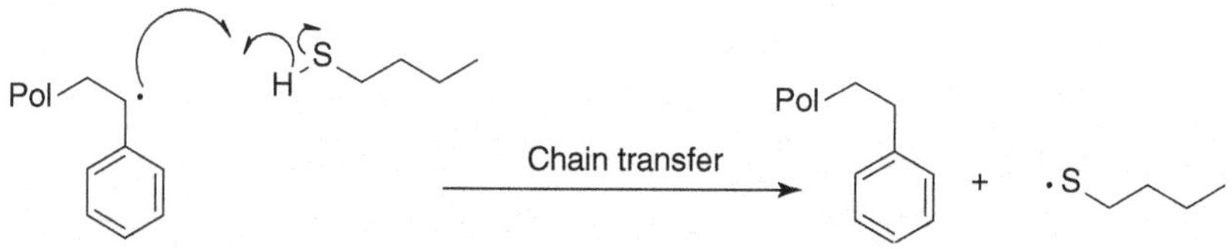

Pol H S Chain transfer Pol + ·S

In an anionic polymerization, the chain transfer agent is anything that can donate a proton. It is illustrated below with isopropanol as the chain transfer agent.

Pol ⊖ H O Chain transfer Pol + ⊖O

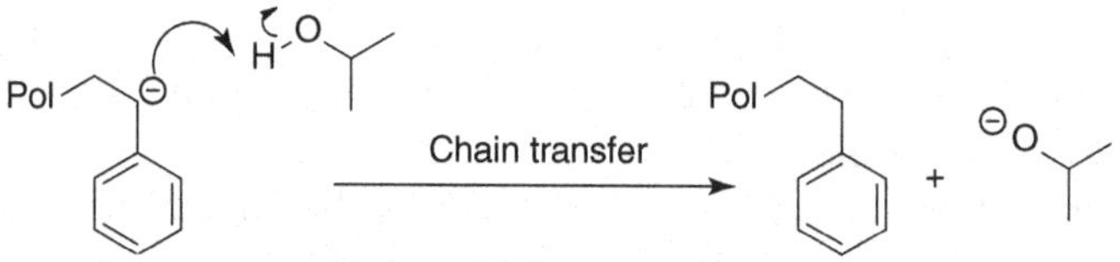

In a living polymerization, there is no chain transfer agent and the polymerization does not terminate. Living polymerizations tend to have a narrower molecular weight distribution and the molecular weight is a function of the amount of added monomer per amount of initiator. For a given amount of initiator, molecular weight increases with increasing amount of monomer. Anionic living polymerizations were discovered in the 1950s when researchers [18] observed that the viscosity of the polystyrene solution continued to increase with addition of more styrene monomer. This enabled anionic block polymers of styrene with rubbers such as butadiene or isoprene. These polymers were commercialized by Shell Chemical Company as Kraton® polymers. For example, styrene can be polymerized in the absence of a chain transfer agent until the styrene monomer is consumed. Butadiene can then be added and a rubber block added to make SB (styrene butadiene) polymer. Then more styrene can be added to the living polymerization to make SBS triblock copolymer. Lastly, the polymerization can be terminated by addition of a chain transfer agent.

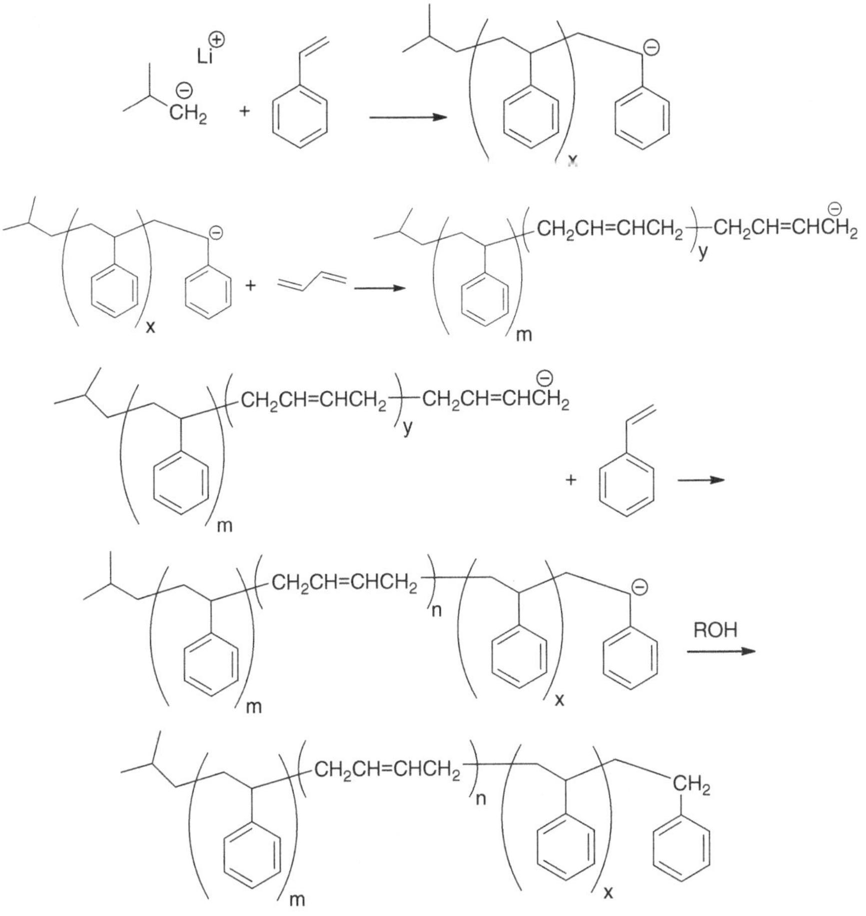

Three decades after the discovery of living anionic polymerizations, living carbocationic polymerizations were discovered [19]. Isobutylene was polymerized in the presence of cumyl acetate and boron trichloride. The molecular weight increased with increasing amount of isobutylene. In a typical cationic polymerization, the reaction is initiated by protonation of the alkene to give a carbocation. Termination happens by elimination to form an alkene.

In a normal (non-living) cationic polymerization, an initiator provides a cation source. This is illustrated below with a proton as the cation. The proton can be formed by the reaction of boron trifluoride catalyst with adventitious water. Isobutylene is protonated to give the tertiary carbocation. This isobutyl carbocation can then react with another molecule of isobutylene to form a new carbocation. This continues and the molecular weight increases. This will continue until either there is no more isobutylene present or until the polymerization is terminated. Termination occurs by loss of a proton to form an alkene. The counterion of the cation can act as a base and promote the termination reaction.

$$BF_3 + H_2O \longrightarrow H^{\oplus} + BF_3OH^{\ominus}$$

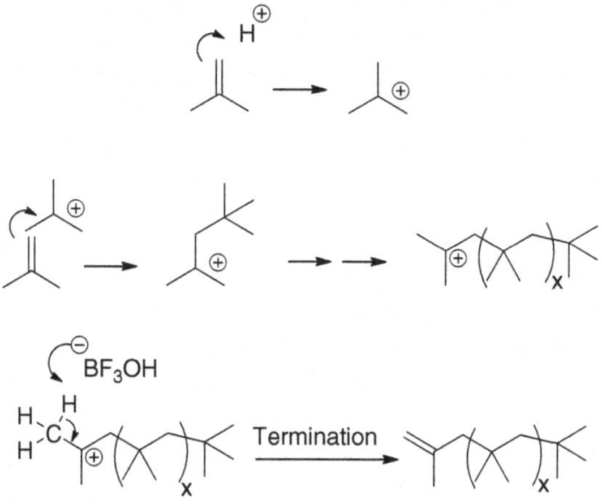

Termination

In a truly living polymerization, there is no termination. However, if the termination step is reversible and if the rate of the reversible reaction is greater than the rate of the polymerization propagation, then we have a quasi-living polymerization [20]. An ideal living polymerization is kinetically indistinguishable from a quasi-living polymerization. In the case of cationic polymerization, it is actually quasi-living. When the original living polymerization of isobutylene was quenched with a nucleophile such as an amine or alcohol, the polymer did not exhibit alkene ends but rather tertiary alkyl chlorides.

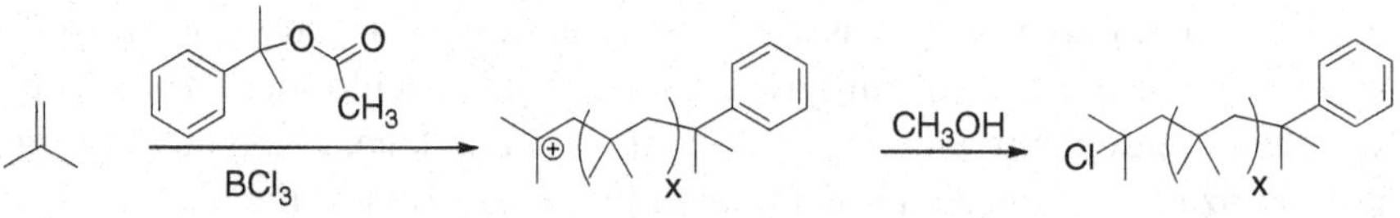

CH_3 BCl_3 CH_3OH Cl

This can be explained if the termination is reversible by the following equilibrium.

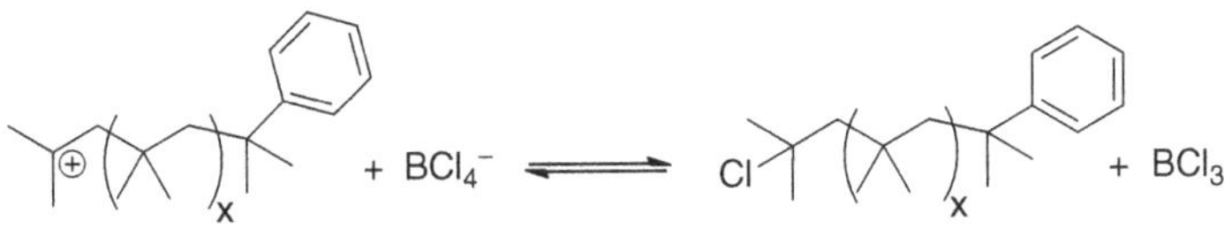

Because of problems with termination due to radical coupling and radical disproportionation reactions, living radical polymerizations were invented much later than living cationic polymerizations which in turn were invented three decades after living anionic polymerizations. Strictly speaking, they are not living because there is not an absence of irreversible termination. Rather, as in the quasi-living cationic polymerizations, there is an equilibrium between the propagating polymer radical and a dormant polymer chain. The term, "reversible-deactivation radical polymerization, RDRP" has been proposed [21] for these radical polymerizations. Different names have been given to these polymerizations depending on the nature of the dormant chain. For example, the dormant chains may be alkyl halides, as in atom transfer radical polymerization (ATRP), thioesters, as in reversible addition fragmentation chain transfer processes (RAFT), or alkoxyamines, asin nitroxide mediated polymerization (NMP) [22].

In a typical ATRP, the dormant chain is terminated as an alkyl halide. The dormant chain interacts with a transition metal in an oxidation/reduction equilibrium (one electron oxidation of the transition metal) to generate the polymer radical which can react with monomer and propagate. The propagating radical reacts with the oxidized transition metal to form a new dormant chain terminated as an alkyl halide. Because the propagating radical exists mainly in the dormant state, there is a relatively low concentration of the active radical and therefore fewer opportunities for termination due to chain coupling or disproportionation.

Perhaps the most common technique is the reversible addition fragmentation chain transfer (RAFT) polymerization [23]. A RAFT polymerization is similar to a typical radical polymerization in that an initiator generates radicals which then combine with monomer to form growing polymer chains. A typical radical polymerization continues until termination occurs either by radical combination or by disproportionation. The length of each chain is statistical and there is typically a mixture of polymeric chains of a broad molecular weight distribution. The distinction is that a RAFT polymerization is carried out in the presence of a chain transfer agent, sometimes called a RAFT agent. Trithiocarbonates are commonly used RAFT agents. A growing polymer chain (Pol·) combines with the RAFT agent. This creates a dormant species which does not undergo termination. It also generates another radical which can react with any available monomer to create a second growing polymer chain (Pol_2·). The second growing chain can combine with the dormant species in an equilibrium. The result of the equilibrium is that all polymer chains grow at a similar rate resulting in a relatively narrow molecular weight distribution. The molecular weight is governed by the amount of monomer and not by chain termination reactions. This polymerization method is especially valuable for preparing block copolymers.

Raft agent
Dormant species
Radical that can reinitiate polymerization with available monomer
Dormant species
Dormant species
Radical that can reinitiate polymerization with available monomer

The molecular weight of a polymer has a pronounced effect on many important physical properties. This can be rationalized by thinking about a plate of spaghetti. The strands are entangled among each other and it can be difficult (and messy) to pull them apart. If the spaghetti represents a high molecular weight polymer, macaroni might represent a low molecular weight polymer. A macaroni noodle is readily removed from the other noodles and is not entangled. In this simplistic analogy, the longer the strand, the greater is the entanglement. Polymers behave similarly and longer strands (higher molecular weight) are more entangled. The greater the entanglement of a polymer, the better is the toughness. Polymer properties such as tensile strength (strength required to pull apart a polymer of a defined cross-sectional area), impact resistance (strength required to fracture a polymer or cause it to yield), and tear strength are all improved (stronger) with increasing molecular weight. However, this is a trade-off with processability. For most plastic fabrication flow in the molten state is necessary and often high flow (low viscosity) is desirable. As a general trend, the greater the polymer molecular weight, the higher the viscosity (and the poorer the flow). Flow is more complicated than this and people have spent their careers studying polymer flow, also known as the rheology of polymers. Polymer molecular weight is a critical parameter for modifying flow.

Because of the pronounced effect of molecular weight on properties, it is well studied. The molecular weight of a monomer is easy to calculate. If we consider ethylene (C_2H_4) we can easily calculate the molecular weight to be 28 amu or 28 g/mol. or 28 daltons. Strictly speaking, these terms are not synonyms, just as weight and mass are not synonyms. However, for our purposes, the distinctions are not critical. The units g/mol are commonly used in the industry and that is what we will use.

If we consider the molecular weight of polyethylene, we should realize that polyethylene consists of many polymer chains of varying length. The molecular weight of a single chain is meaningless; rather we need to consider the average of the large number of chains. One way to think about the average molecular weight is to add up the molecular weight of each polymer chain and then divide by the number of chains. This is the number average molecular weight, symbolized as M_n. All polymer chains do not have an equal importance and larger molecular weight chains can have a greater influence than smaller molecular weight chains. Therefore, a weighted average is often considered.

An analogy might be useful. If a 50 foot Beneteau sailboat weighs 14,000 kg and after sitting in the water, 999 barnacles, each weighing 1 g (or 0.001 kg), grow on the hull, then what is the average weight? The number average is 14 kg, but this isn't particularly descriptive of the system. The weighted average (for a polymer, this would be the weight average molecular weight or M_w) is 13,999 kg, which is likely to be more descriptive of the barnacle-ridden Beneteau. Similarly the weight average molecular weight is more relevant for many polymer applications.

To illustrate the arithmetic, let's pick an example of a small toy boat weighing 100.00 g and three barnacles, each weighing 1.00 g. The number average is the sum of the masses divided by the number of items. In this case, it is a total weight of 103.0 g divided by 4 which gives us 25.75 g as the number average weight. To calculate the weighted average, multiply each weight by itself divided by the total weight and add the products. This gives the weighted average of 97.12g.

$$
\begin{aligned}
100.00 \times \frac{100.00}{103.00} &= 97.09 \\
1.00 \times \frac{1.00}{103.00} &= 0.01 \\
1.00 \times \frac{1.00}{103.00} &= 0.01 \\
1.00 \times \frac{1.00}{103.00} &= 0.01 \\
\hline
&\quad 97.12
\end{aligned}
$$

As another example, let's consider two polymer chains, one with a molecular weight of 40,000 g/mol and the other with a molecular weight of

80,000 g/mol. The polymer M_n is (40,000 + 80,000)/2 = 60,000 g/mol. The M_w is 66,666 g/mol.

$$40{,}000 \times \frac{40{,}000}{120{,}000} = 13{,}333$$

$$80{,}000 \times \frac{80{,}000}{120{,}000} = 53{,}333$$

$$\overline{66{,}666}$$

Another common term is polydispersity, which is the ratio of M_w/M_n. In the above example, the polydispersity is (66,666/60,000) = 1.11. If each polymer chain in a sample had the exact same molecular weight, then the polydispersity would be 1.0. As the spread of molecular weight among the various polymer chains increases, the polydispersity increases. A uniform process with a catalyst that produces polymer chains of similar molecular weight might have a reasonably low polydispersity such as 2, and another system, perhaps with a catalyst that has sites of different activities, might give a broader distribution of molecular weights and have polydispersity of, for example, 5 or 6. A blend of polymers with remarkably different molecular weights might have a polydispersity of greater than 15.

Two polymer samples can have the same weighted average but very different polydispersity. In this case, many properties will be different because both the molecular weight and the polydispersity have a strong influence on properties.

The degree of polymerization of a polymer is the number of repeat units in that polymer. A particular polyethylene chain with a molecular weight of 28,000 g/mol has a degree of polymerization of 1,000. This is calculated by dividing 28,000 by the repeat unit weight of 28 (C_2H_4). For this example, n = 1,000 in the structure below.

$$\left(\begin{array}{cc} H & H \\ | & | \\ -C- & -C- \\ | & | \\ H & H \end{array} \right)_n$$

Because molecular weight is so important in determining physical properties such as impact and strength, it is routinely measured in the polymer industry. One convenient method to measure molecular weight is by Gel Permeation Chromatography (GPC). This method has some similarity to HPLC in that a solution of the analyte (in this case the polymer) is passed through a column at high pressures and then as it exits the column detected by various methods such as UV, refractive index (RI), or light scattering. In HPLC, the retention time is determined by affinity of the analyte toward the packing material of the column. In normal phase HPLC, analytes are separated based upon their affinity for a polar column material such as silica. In reversed-phase HPLC, non-polar packing material is used as the stationary phase. In contrast, GPC separates materials based upon molecular size. The stationary phase has pores. Smaller molecules can enter the pores and therefore take longer to elute from the column. Larger molecules are too big to enter the pores; they are excluded from the pores. Therefore, larger molecules flow faster through the column. Because of this principle, GPC is sometimes called size exclusion chromatography. The shorter the retention time, the larger the molecule. Commonly more than one column is used with the different columns using stationary phases of different pore sizes. By doing this, distinctions can be made between polymer chains of various molecular weight. Standards of known molecular weight can be analyzed by GPC and a calibration curve developed that correlates retention time to molecular weight. Polystyrene standards are readily available and are often used, even when analyzing other types of polymers. The GPC can give M_n, M_w and polydispersity data.

Sometimes other techniques are used to gain information about molecular weight. Because molecular weight has such a profound influence on rheological properties, sometimes the viscosity is analyzed and then inferences are made about molecular weight. The tests can be very simple. For example, a polymer can be placed in a heated cylinder and, after allowing time for melting, forced through a die by placing a weighted rod on top of the cylinder. The weight pushes the rod which in turn extrudes the polymer melt through the die. By weighing the amount of extrudate per minute, you can have a measure of viscosity. The greater the weight of extrudate, the less viscous the material. After multiplying by ten, you have the melt index which is the grams extrudate per ten minutes. The melt index is also called the melt flow index and is an easy quality control test.

Another technique which gives rheological information is Dynamic Mechanical Analysis (DMA), also known as Dynamic Mechanical Thermal Analysis (DMTA). Small deformations are applied to a polymer melt in a cyclic manner. The polymer melt can be held between two parallel plates which oscillate. This technique is much more sophisticated than a melt index and gives much more information. The force on the melt can be varied and the effect of force on flow (called shear sensitivity) can be measured. Some polymers, especially those with high molecular weight or some branching, have high shear sensitivity. To understand shear sensitivity, think about water which flows easily and then toothpaste. Toothpaste doesn't flow unless a force (shear) is applied. Then, when you squeeze the tube, toothpaste flows readily. Toothpaste has high shear sensitivity.

For some polymers, portions of their chains can assemble with portions of other polymer chains and form highly ordered crystalline regions. This is the same as when a small molecule forms crystals. Like small molecules, crystalline polymers have melting points (often abbreviated as T_m). However, unlike small molecules, polymers are never completely crystalline. Because the regions that can pack together are present within a chain and because of chain entanglement, there is a limit to how much of the polymer chain can be ordered and be in crystal form. A crystalline polymer is never 100% crystalline and depending on the polymer, values of 50–95% crystallinity are common. The T_m of a crystalline polymer determines its maximum upper-use temperature. Above the T_m, the polymer softens and does not retain its physical properties.

Not all polymers are crystalline. For these amorphous polymers it is the glass transition temperature (T_g) that determines the upper-use temperature. An amorphous polymer is said to be in the hard, glassy state when it is at a temperature below the T_g. At this temperature, there is only localized motion of the polymer. Bonds are stretching and wagging, but the molecules remain in place. As the temperature is increased, the thermal energy is transferred to kinetic energy and the motion increases. Concurrent with the increase in localized motion, is an increase in volume. This is due to vibrational motion, the bending, stretching, and wagging of bonds. As the temperature continues to increase, at a certain temperature, the T_g, there is enough energy to enable longer range motion of the molecules. They are no longer "frozen" in place but can slide by each other. Above the T_g, the polymer becomes soft. The volume continues to increase with increasing temperature, but at a different rate. The T_g defines the point at which there is a change in slope of the volume/temperature curve. A crystalline polymer has both a T_g and a T_m. The T_g is at a lower temperaturebut it is thc $T_m$that determines the upper use temperature.

QUESTIONS

1. If styrene costs 42.0 ¢ per pound and the price of the catalyst is negligible, assuming 100% yield calculate the raw material cost of polystyrene to the nearest tenth of a cent.

2. Label (name) each of the following to indicate the stereochemistry.

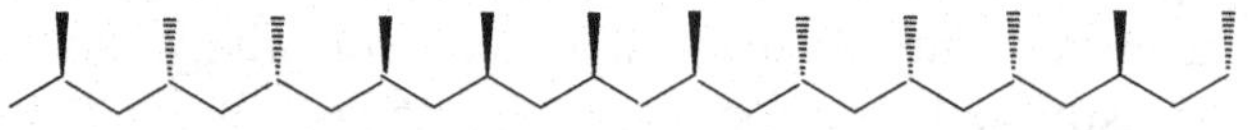

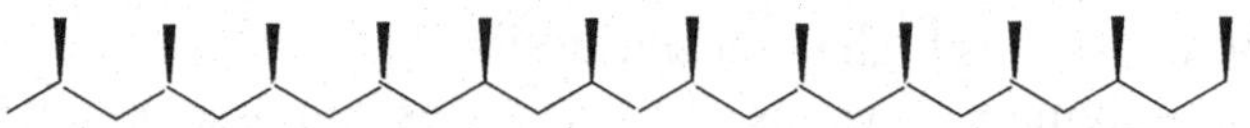

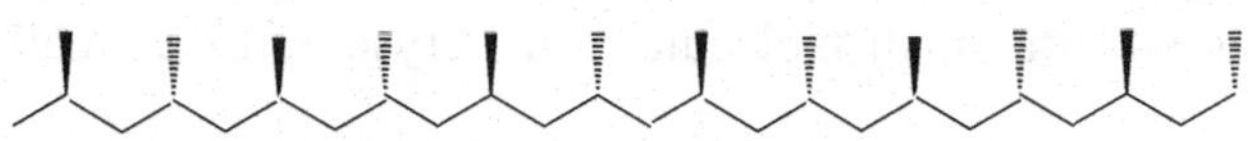

3. Write the mechanism using curved arrows and showing electron flow for the anionic polymerization of styrene using butyl lithium.

4. Consider the following four polymers and select any that have sufficient thermal properties to be used to fabricate dishware to be used in a dishwasher with a steam cycle. (the steam is not pressurized; select all that apply; could be any number from 0 to 4 choices).
   a. an amorphous polymer with a Tg of 60 °C
   b. a semicrystalline polymer with a Tg of 40 °C and a melting point of 220 °C
   c. an amorphous polymer with a Tg of 150 °C
   d. a semicrystalline polymer with a Tg of −20 °C and a melting point of 150 °C

5. Select the pair that won the 1963 Nobel prize for their work in olefin polymerization catalyzed by organometallics such as $TiCl_3/Et_3Al/MgCl_2$
   a. Simon and Garfunkel
   b. Joule and Rutherford
   c. Ziegler and Natta
   d. Abbott and Costello
   e. Martin & Lewis

6. If we imagine a group of three polymer molecules with molecular weights of 40,000 g/mol, 180,000 g/mol, and 200,000 g/mol, calculate (to three significant figures) M_n, M_w and polydispersity (Pd).

7. If I am driving a Kia with a polypropylene bumper, which polypropylene would be best if I were in a collision on College Road?
   a. polypropylene with the highest M_w
   b. polypropylene with the lowest M_w
   c. polypropylene with the lowest viscosity
   d. polypropylene with the highest Tg.

8. Although, perhaps not an oxymoron (e.g. jumbo shrimp), which of the following word pairs indicates a lack of understanding about polymers.
   a. syndiotactic polyethylene
   b. atactic polystyrene
   c. isotactic polypropylene
   d. atactic polypropylene

9. Consider the two superimposed GPC curves. On the x axis, is the retention time in minutes (from 5 minutes to 14 minutes). Sample A begins eluting around 6 minutes and has all eluted by about 11 minutes; Sample B begins eluting around 8 minutes.

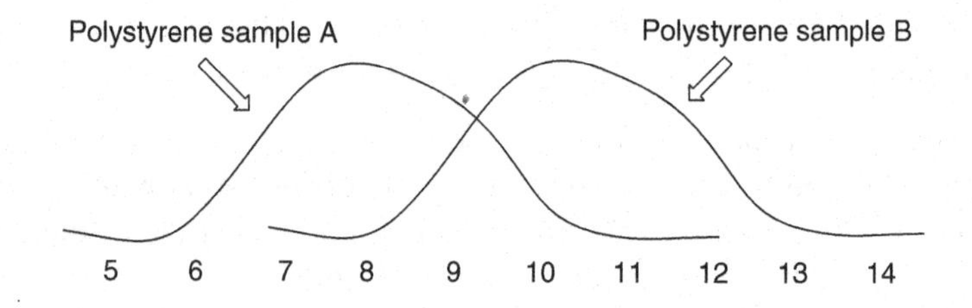

Which is likely to have the greater tensile strength?

a. Sample A

b. Sample B

c. they should have the same

d. it's a coin flip; the GPC doesn't give any pertinent information

10. Consider three polyethylene chains, one with molecular weight 60,000, one with molecular weight 100,000, and one with molecular weight 200,000.

a. to the nearest 1,000, what is the M_n?

b. to the nearest 1,000, what is the M_w?

c. to three significant figures, what is the polydispersity?

11. Generally as molecular weight increases:

a. melt viscosity and impact resistance increase

b. melt viscosity and impact resistance decrease

c. melt viscosity increases and impact resistance decreases

d. melt viscosity decreases and impact resistance increases

REFERENCES

1. Graeme Moad, Ezio Rizzardo, David H Solomon. Macromolecules 1982: 15:909–914.
2. Yoshiyuki Ishihama, Mitsuo Sawamoto, Toshinobu Higashimura. Polymer Bulletin 1990; 24:201–206.
3. YV Kissin. Journal of Catalysis 2012; 292:188–200.
4. JP Hogan, RL Banks. U.S. Pat. No. 2,825,721. 1958.
5. A Andresen, HG Cordes, J Herwig, W Kaminsky, A Merck, R Mottweiler, J Pein, H Sinn, HJ Vollmer. Angewandte Chemie 1976; 88(20): 689–690.
6. TJ Kealy, PL Pauson. Nature 1951; 168:1039–1040.
7. SA Miller,JA Tebboth, JF Tremaine. J. Chem. Soc. 1952:632–635.
8. G Wilkinson, M Rosenblum, MC Whiting, RB Woodward, J. Am.Chem Soc. 1952; 74:2125–2126.

9. Pierre Laszlo, Roald Hoffmann. Angew. Chem. Int. Ed. 2000; 39(1):123–124.
10. M Ystenes, JL Eilertsen, J Liu, M Ott, E Rytter, JA Stovneng. Journal of Polymer Science: Part A: Polymer Chemistry 2000; 38:3106–3127.
11. HG Alt, A Koppl. Chem. Rev. 2000:1205–1221.
12. DR Neithamer, JC Stevens. U.S. Pat. No. 5,399,635. 1995.
13. S Nagy, BP Etherton, R Krishnamurti, JA Tyrell. U.S. Pat. No. 6,376,629. 2002.
14. S Nagy, BP Etherton, R Krishnamurti. U.S. Pat. No. 6,034,027. 2000.
15. R Schrock. Acc. Chem. Res. 1990; 23:158–165.
16. M Piotti. Current Opinions in Solid State and Materials Science 1999; 4:539–547.
17. C Bielawski, R Grubbs. Prog. Polym. Sci. 2007; 32:1–29.
18. M Szwarc, M Levy, R Milkovich. J. Am. Chem. Soc. 1956; 78:2656–2657.
19. R Faust, J Kennedy. Polymer Bulletin 1986; 15:317–323.
20. J Kennedy. Journal of Polymer Science: Part A: Polymer Chemistry 1999; 37:2285–2293.
21. AD Jenkins, RG Jones, G Moad. Pure Appl. Chem. 2010; 82:483–491.
22. K Matyjaszewski, J Xia. Chem. Rev. 2001; 101:2921–2990.
23. G Moad, E Rizzardo, San Thang. Acc. Chem. Res. 2008; 41:1133–1142.

CHAPTER 8

Some Industrially Important Polymers

8.1 POLYETHYLENE

Polyethylene is the largest volume polymer produced and its precursor ethylene is the largest volume organic chemical produced. The reason that polyethylene is so widely used is that it has an attractive balance of properties at an economical price. Polyethylene can be fabricated by a variety of techniques such as injection molding, blow molding and rotomolding. It has excellent chemical resistance and strength. The end use of a plastic is an article – a bag, bottle, film, bumper, computer housing, medical implant, and so forth. The polymer to make that article is chosen based upon its physical properties and ability to be formed into the article. If these criteria are met by several polymers, then cost usually determines which polymer is used. Polyethylene is the lowest cost polymer for a variety of reasons. The raw material, ethylene, is low cost. Because polyethylene is made in such high volumes, the manufacturing cost is low. Because the density is relatively low, it takes less weight to make an article than if a higher density material were used. To emphasize this last point, consider an article that could be made from either polyethylene having a density of 0.92 g/cm^3 or polycarbonate having a density of 1.20 g/cm^3. These polymers are purchased by weight, so many cents or dollars per pound. Even if the purchase price were the same (and it is not – polyethylene is much lower), the material cost to make the polyethylene article is about 23% lower.

Polyethylene was originally made by a high pressure, high temperature radical process. Reactor pressures may be as high as 40,000 psig and temperatures as high as 300 °C. Because of the indiscriminant nature of radical reactions done at these high temperatures the polymer is not linear. It has oligomeric branches as shown below. Note that the polymer has many more repeat units and this is a sketch of only a section of the molecule.

Fundamentals of Industrial Chemistry: Pharmaceuticals, Polymers, and Business, First Edition. John A. Tyrell.

This reduces the ability of the polyethylene to pack closely together and therefore lowers the density. Polyethylene of this type is called low density polyethylene (LDPE) and has a density in the approximate range of 0.910–0.930. The structure gives LDPE certain property advantages, especially regarding rheological properties. LDPE is made to this day, but now most polyethylene is made with organometallic catalysts and is more linear.

Polyethylene made by organometallic catalysts such as Ziegler–Natta or metallocene systems is linear and can be represented as follows.

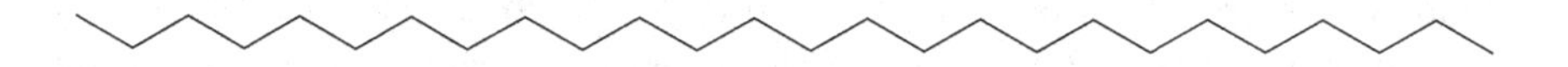

A linear ethylene homopolymer can pack together more tightly and has a higher density, somewhere around 0.95 g/cm^3. High density polyethylene (HDPE) is generally considered to have a density >0.940 g/cm^3. The density of ethylene homopolymer can be decreased by copolymerizing an alpha-olefin with the ethylene. Major commercial polyethylene copolymers are based upon using 1-butene, 1-hexene, or 1-octene. Of these three, the lowest cost comonomer is 1-butene; 1-octene is the highest cost and has the greatest effect. Many producers use 1-hexene which is a compromise between cost and performance. The copolymer of ethylene with 1-hexene can be represented as follows.

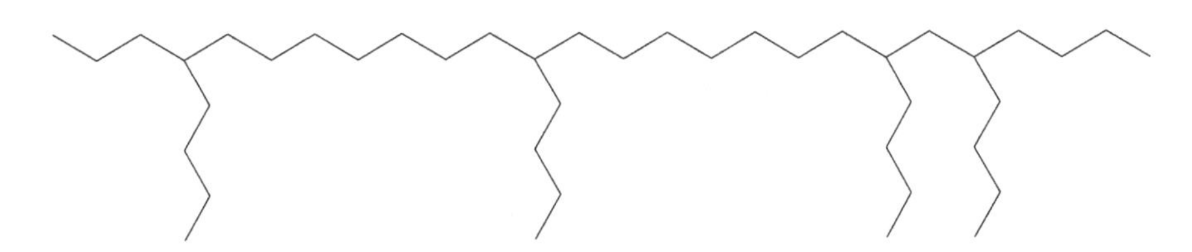

The result is a linear polymer with branches that are four carbons in length. The branches decrease the ability of the polyolefin chains to pack tightly and decrease the density. The greater the number of branches, the lower is the density. Depending on the process and the catalyst system, the comonomer can incorporate somewhat uniformly. However, in other instances the comonomer incorporation is less uniform and can be more prevalent in lower molecular weight chains. In the extreme, this can result in a lower melting waxy material which can give process problems. The choice of comonomer, the amount of comonomer, and the uniformity of incorporation all influence properties and differentiate various grades of polyethylene.

HDPE has small levels of or no comonomer and a density of 0.940–0.970 g/cm^3. Linear low density polyethylene (LLDPE) has a greater amount of comonomer. The grade cutoffs can vary from company to company, but generally linear polyethylene with a density from 0.890–0.920 g/cm^3 is considered LLDPE and 0.920–0.940 g/cm^3 is medium density polyethylene (MDPE). It may seem surprising that such distinctions are made among seemingly small changes in density, but these changes are significant. In the industry, densities are routinely measured to four significant figures. As a general trend, as density increases, crystallinity increases. Modulus (stiffness) and heat softening points also increase with increasing density. Permeation decreases with increasing density. Low permeation is usually desirable so a polyethylene pipe for natural gas would have a lower rate of natural gas permeation if it were made from HDPE than if from LLDPE. Generally, low temperature toughness decreases with increasing density. A polyethylene tub on a cold winter day in Buffalo left outside and then dropped is more likely to break if it is made from HDPE than if made with LLDPE.

Another differentiating property of polyethylene is molecular weight. The molecular weight is influenced by the choice of catalyst system and the polymerization conditions. As the polymerization proceeds, the active site can decompose in a chain termination reaction. The molecular weight is determined by the relative reaction rates of the propagation and termination reactions. Many catalyst systems make very high molecular weights, and in these circumstances a chain termination agent, usually hydrogen, is added to control molecular weight. The polydispersity can be influenced by the nature of the catalyst system. A single-site catalyst will give a narrower polydispersity versus a Ziegler–Natta system. Combinations of catalysts can be used to increase M_w/M_n. The use of multiple reactors in a polymerization process also allows for changing the polydispersity by varying the conditions in each reactor.

Polyethylene is made in a continuous fashion by solution, slurry, and gas-phase processes. In a solution process, the reaction is run at high temperatures (maybe 200–300 °C) in the presence of an organic solvent such as hexane. At the reaction temperature, the ethylene and the polyethylene remain in solution during the polymerization. After the reaction, the pressure is released and the solvent removed. In a slurry process, the temperature is lower and the polyethylene forms as a slurry in the organic solvent. In a gas-phase process, solid polyethylene seeds are suspended in a gaseous stream of ethylene and the ethylene polymerizes onto these suspended seeds. These descriptions are gross oversimplifications and much process research has been, and continues to be, done in this area. Details of each of these processes and recent improvements can be found by consulting the patent literature.

Polyethylene can be processed by a variety of techniques. It can be blown into film which is then used to make items such as shrink wrap and the ubiquitous grocery bags. It can be injection molded into a variety of shapes or blow molded into milk bottles and detergent bottles. It can be rotomolded into large parts such as toys, agricultural tanks, or kayaks.

Polyethylene has limited adhesion to paints and inks. This is because it is a non-polar hydrocarbon incapable of forming hydrogen bonds. Adhesion can be improved by copolymerizing ethylene with polar monomers such as ethyl acrylate or vinyl acetate to give ethylene ethyl acrylate (EEA) copolymers or ethylene vinyl acetate (EVA) copolymers. EVA is often used for shoe soles.

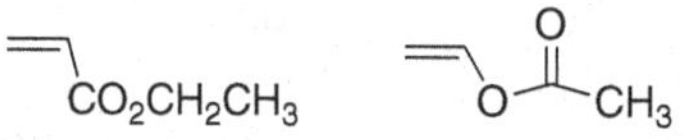

$CO_2CH_2CH_3$ — Ethyl acrylate

O, O, CH_3 — Vinyl acetate

8.2 POLYPROPYLENE

Polypropylene has many of the same desirable properties of polyethylene, but has a higher heat capability and excellent fatigue properties. Most commercial polypropylene is made with Ziegler–Natta catalysts in a gas-phase process and is isotactic. The T_m is about 160 °C, about 50 °C higher than HDPE. Because of this it can be used for applications requiring somewhat higher temperatures such as those that need sterilization by autoclave.

Unlike ethylene, propylene cannot be polymerized to a high molecular weight by a radical process [1]. Competing with the propagation step is hydrogen atom abstraction from propylene which gives an allyl radical and a terminated oligomer.

$$R{-}CH_2{-}\dot{C}H(CH_3) + H_2C{=}CH(CH_3) \longrightarrow R{-}CH_2{-}CH(CH_3){-}CH_2{-}\dot{C}H(CH_3) \longrightarrow \longrightarrow$$

$$R{-}CH_2{-}\dot{C}H(CH_3) + H{-}CH_2{-}CH{=}CH_2 \longrightarrow R{-}CH_2{-}CH(CH_3){-}H + \cdot CH_2{-}CH{=}CH_2$$

Allyl radical

Polypropylene has excellent fatigue resistance. Fatigue is the process by which materials fail after cyclic loading. Consider a piece of bare copper wire. It is strong and cannot be separated by hand. However, if you bend it back and forth several times, eventually it will fail and break apart. This is fatigue failure. Polypropylene is very resistant to fatigue. Therefore it is used in applications, such as plastic hinges, which require repeated flexing, bending, or opening and closing. Other polymers develop cracks and break upon repeated bending.

Polypropylene makes good fibers and is used for clothing, ropes, and carpets. Unlike wool or nylon, polypropylene does not soak up water so it is used for outdoor carpeting and athletic clothing. Other applications for polypropylene include yogurt cups, straws, bottle caps, and automobile bumpers.

8.3 POLYVINYL CHLORIDE

Although vinyl chloride has been polymerized with organometallics [2], commercial poly(vinyl chloride) (PVC) is made by a free radical polymerization. The dominant process is the suspension process. Vinyl chloride (most commonly referred to as vinyl chloride monomer or VCM) is suspended as droplets in water and an initiator that is soluble in the VCM is added. VCM has a boiling point of $-13\,^{\circ}C$. It is reacted under pressure as a liquid. The resultant PVC powder is separated from the water and dried. PVC can also be prepared by a mass or bulk polymerization where the initiator is added to the liquid VCM. A third polymerization technique, an emulsion polymerization, employs liquid VCM, water, and an emulsifier. A water soluble initiator is added to the VCM/water emulsion.

The PVC which is formed in the reactor has the tendency to adhere to the walls of the reactor, the agitator, and reactor baffles. If it later flakes off, it contaminates subsequent batches and can cause defects. The flakes are often less soluble and darker. These imperfections are particularly noticeable in film applications where they cause optical defects due to gels. These film defects are called "fisheyes." To minimize this, reactors are periodically cleaned. For this reason, most PVC processes are done batchwise or semi-continuous and not in true continuous fashion like polyethylene or polypropylene. At one time, PVC adhering to polymer walls was such a problem that people had to go inside the reactors and manually clean them. Process improvements have since been made including a discovery about coating of reactor surfaces to minimize adhesion [3].

In the case of VCM, the head of the monomer unit is arbitrarily designated as the chlorine-containing carbon and the tail as the non-chlorine containing carbon [4]. Because the radical intermediate is more stable when the carbon has a chlorine atom, the monomers assemble in an almost exclusively head to tail fashion.

Head to tail
more stable radical formed

Head to head
less stable radical

Although most of the PVC is an assemblage of VCM in a head to tail fashion, there are different types of structural defects which have been found in PVC. Because of these defects, PVC has poorer thermal stability than would be otherwise expected. Thermal degradation occurs at one of these weak spots and the polymer can eliminate HCl from the neighboring two carbons. The resultant olefin destabilizes the neighboring bonds and another HCl molecule is eliminated, with eventually, HCl continuing to "unzipper" all along the polymer chain. Two defects thought to cause much of this degradation are allylic chlorines and chlorines on tertiary carbons [5].

Because of this tendency towards HCl elimination, stabilizers are necessary for PVC. There has been a large amount of research in this area. Lead stabilizers are effective but are no longer used to any extent in the U.S. When some imported toys are found to have lead contamination, if it is not from the paint or pigments, it can be from PVC made with lead stabilizers. For some time in the U.S., tin-based stabilizers have been widely used. Compounds such as dibutyltin bis (isooctylthioglycolate) are effective.

PVC is used in a variety of applications from rigid applications such as pipe and vinyl siding to flexible applications such as black electrical tape, garden hoses, and tubing. The flexibility is imparted by plasticizers. A plasticizer is an additive that reduces intermolecular forces in polymers and makes the polymer more flexible [6]. More than 12 billion pounds of plasticizers are sold globally each year, with 96% used to impart flexibility to PVC [7]. There are many types of plasticizers, but commonly they are esters such as esters of adipic acid, trimellitic acid, or phthalic acid.

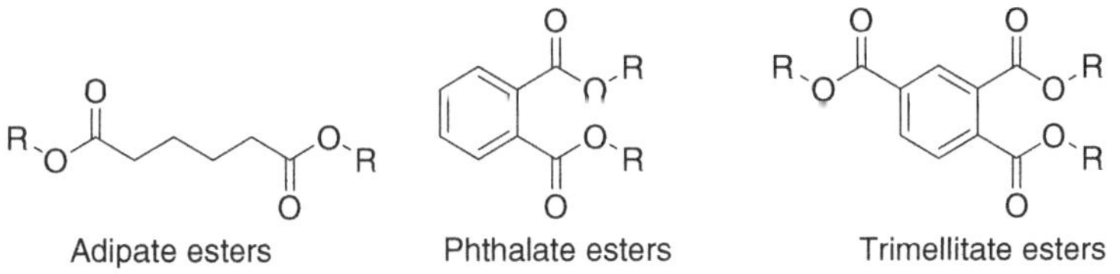

Adipate esters   Phthalate esters   Trimellitate esters

Dioctyl phthalate (DOP), more properly called di-2-ethylhexyl phthalate (DEHP), is commonly used as an inexpensive general purpose plasticizer.

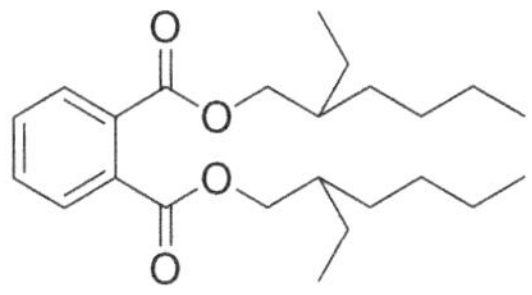

DOP

These plasticizers have been used for more than 50 years in applications such as IV bags and other medical devices. However, there is some uncertainty and debate within the scientific community about possible health effects, especially with DOP. To date, no studies have evaluated the potential for DOP to cause cancer in humans. Eating high doses of DOP for a long time resulted in liver cancer in rats and mice. The U.S. Department of Health and Human Services has determined that DOP may reasonably be anticipated to be a human carcinogen. EPA has determined that DOP is a probable human carcinogen. These determinations were based entirely on liver cancer in rats and mice. The International Agency for Research on Cancer (IARC) has recently changed its classification for DOP from "possibly carcinogenic to humans" to "cannot be classified as to its carcinogenicity to humans," because of the differences in how the livers of humans and primates respond to DOP as compared with the livers of rats and mice [8].

The choice of plasticizer is based upon many factors and depends upon the end-use application. Cost, flexibility imparted per pound of plasticizer, low temperature performance, volatility, and other factors are commonly important in the selection process.

The ability to absorb a plasticizer is important for PVC. Porosity of the PVC particle is an important quality parameter of PVC and is determined by the polymerization process. Another characteristic of PVC that is determined by the polymerization is particle size, with emulsion processes giving smaller particles than suspension processes.

8.4 OTHER OLEFIN POLYMERS

There are many other commercially important olefin polymers based upon polymerization of available substituted olefins.

$$H_2C{=}CXY \longrightarrow \left(CH_2{-}CXY \right)_m$$

For example, when X is hydrogen and Y is phenyl, we have the polymerization of styrene. Styrene can be polymerized by a variety of processes including cationic, anionic, and organometallic. Metallocene catalysts have been used to prepare syndiotactic polystyrene which is crystalline and has improved heat capability. Most commercial polystyrene is made by a radical polymerization process and is atactic. It is a low cost polymer and is useful for its clarity, stiffness, and ability to be foamed. As a homopolymer, sometimes referred to as general purpose (GP) polystyrene, it lacks toughness. It is used for CD and DVD cases, disposable clear drinking cups, and plastic utensils. It can be expanded or foamed to make disposable coffee cups, inexpensive coolers, and insulation. Dow trademarked this as styrofoam®. The problem with toughness is solved by the addition of rubber particles such as polybutadiene to make high impact polystyrene (HIPS). The rubber particles can be dispersed in styrene and the styrene then polymerized. This method grafts some of the polybutadiene to the styrene [9, 10].

One common terpolymer containing styrene is acrylonitrile-butadiene-styrene, known as ABS. The acrylonitrile imparts heat stability and solvent resistance. The butadiene imparts impact strength and the styrene imparts stiffness, also known as high modulus, and good processability.

When X is hydrogen and Y is a carboalkoxy group, we polymerize to make acrylates. If X is methyl and Y is carbomethoxy, we have methyl methacrylate polymers. These are more commonly known by their tradenames: Perspex®, ICI; Plexiglas, Rohm & Haas; and Lucite, Dupont.

$$H_2C{=}C(CH_3)(CO_2CH_3) \longrightarrow \left(CH_2{-}C(CH_3)(CO_2CH_3) \right)_m$$

Polymethyl methacrylate

Many of the acrylates are also made by free radical polymerization. By varying the alcohol portion, such as methanol as in methyl acrylate or butanol for butyl acrylate or 2-ethylhexanol as in octyl acrylate, the Tg and other polymeric properties can be varied. As the side chain increases in length, the T_g decreases. Going from a one carbon alcohol as in methyl acrylate to an eight carbon alcohol results in a lowering of the Tg by about 60 °C. However, the cost increases. Often copolymers are made such as in the copolymerization of methyl acrylate with butyl acrylate to optimize properties needed for particular applications. The acrylates are commonly used as pressure-sensitive adhesives.

If X is cyano and Y is carboalkoxy we have the cyanoacrylate polymers, useful as adhesives. For example, ethyl cyanoacrylate is marketed as Super Glue® or Krazy Glue®.

$$H_2C{=}C(CN)CO_2CH_2CH_3 \longrightarrow \left(CH_2{-}C(CN)(CO_2CH_2CH_3) \right)_m$$

Ethyl cyanoacrylate

The cyanoacrylate olefins have two electron-withdrawing groups on the substituted carbon of the olefin and therefore can be readily polymerized in anionic fashion by use of a base. For example, N,N-dialkyl anilines have been used to activate the polymerization [11].

8.5 POLYESTER

An ester can be prepared by the reaction of a carboxylic acid and an alcohol as illustrated with benzoic acid and ethanol to make ethyl benzoate. Typically, the reaction is done with an acid catalyst, such as sulfuric acid, and the equilibrium is shifted to the right either by removal of the water or by using a large excess of the alcohol or both. In the reaction, an equivalent of water is condensed or removed for each ester bond formed.

$$C_6H_5COOH + CH_3CH_2OH \rightleftharpoons C_6H_5COOCH_2CH_3 + H_2O$$

The reaction can also be done by a transesterification reaction. This can be illustrated by the reaction between methyl benzoate and ethanol to form ethyl benzoate and methanol. This too is acid catalyzed. The equilibrium can be shifted to the right by using an excess of ethanol or by removing methanol as it forms. This is also considered a condensation reaction because methanol is removed by condensation of the methyl benzoate with ethanol.

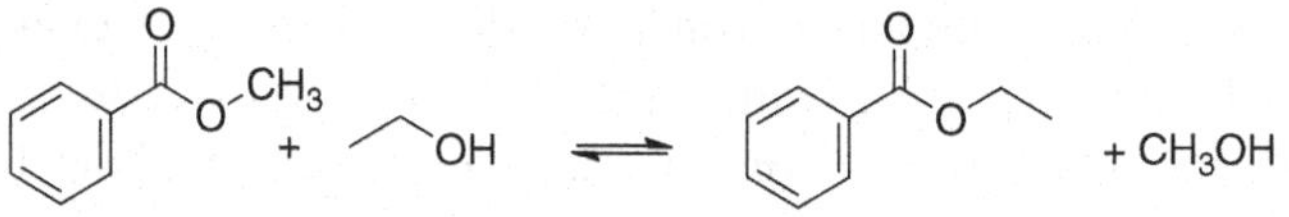

+ OH ⇌ + CH_3OH

These reactions are the basis for the commercial preparation of polyesters. Polyesters are formed from a dicarboxylic acid and a dialcohol. The largest volume polyester is polyethylene terephthalate (PET). Rynite®(Dupont) is a common brand. This is a condensation polymer and made by the removal of water in the condensation reaction of ethylene glycol with terephthalic acid. Note that all of the atoms of the monomers are not present in the final polymer. This is different from addition polymers.

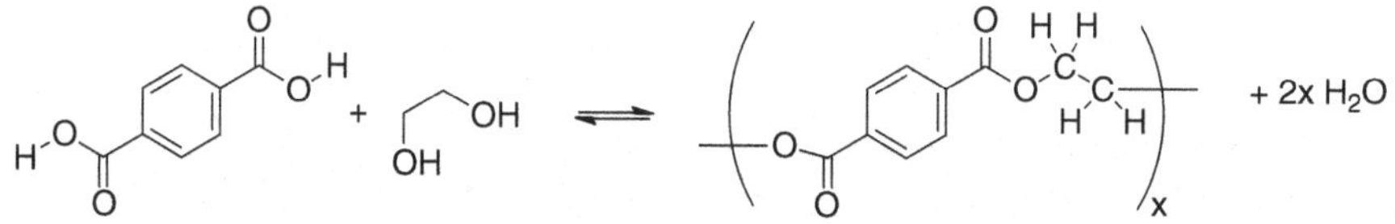

+ 2x H_2O

Both terephthalic acid and ethylene glycol are prepared by oxidation of chemicals isolated from petroleum and are therefore economically attractive.

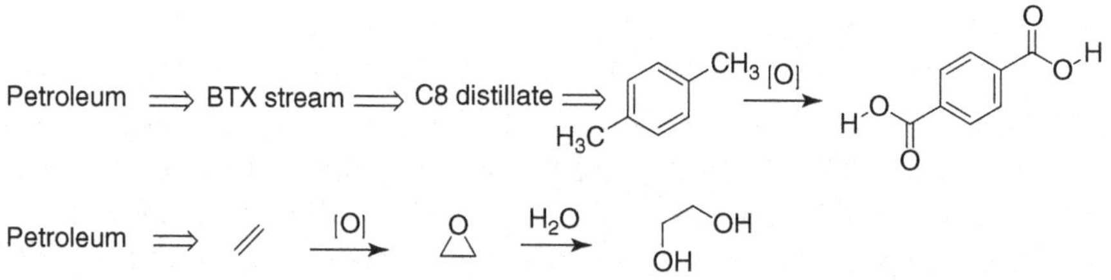

Petroleum ⟹ BTX stream ⟹ C8 distillate ⟹ [O] →

Petroleum ⟹ [O] → H_2O →

The polymerization is catalyzed by acid and the equilibrium is driven to the right by the removal of water. However, there is another competing reaction, etherification, which is also acid catalyzed and driven by the removal of water.

$$2\ ROH \rightleftharpoons ROR + H_2O$$

Ether formation

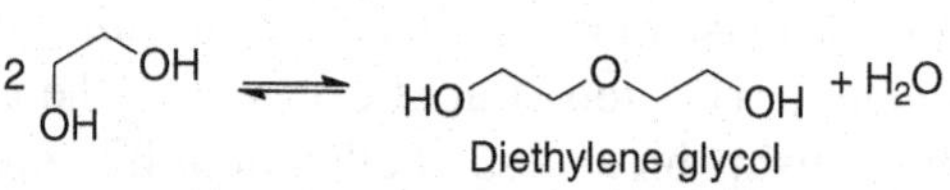

2 ⇌ HO–O–OH + H_2O

Diethylene glycol

In the polymerization to give PET, any diethylene glycol that is formed will react, just as ethylene glycol does, and incorporate into the polymer chain. PET is a crystalline polymer and its melting point of about 260 °C determines its upper use temperature. If a significant amount of diethylene glycol is formed, it will incorporate into the polyester and lower the melting point. It will also decrease the crystallinity of the PET. Therefore, instead of using sulfuric acid the acid catalyst is typically a Lewis acid that enables the esterification reaction to preferentially occur. Antimony oxide is commonly used.

Polyesters are also made by transesterification reactions. Terephthalic acid reacts with methanol to give dimethyl terephthalate (DMT) which then reacts with ethylene glycol to give PET. Again a Lewis acid is used as a transesterification catalyst. Common catalysts are tetraalkyl titanates or antimony oxide.

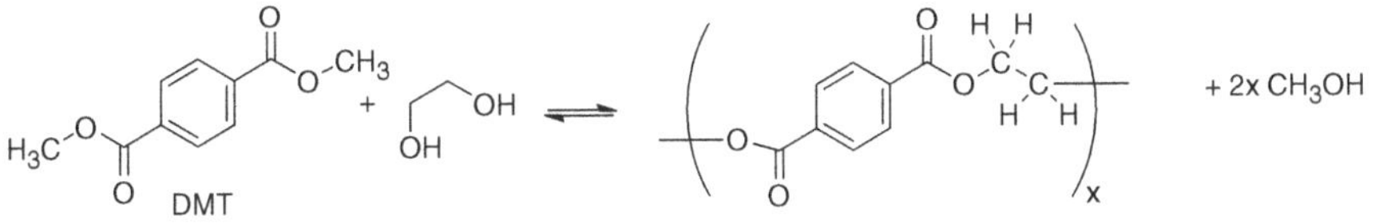

Although conversion of terephthalic acid to DMT adds an extra step, there are some advantages that make it attractive. Terephthalic acid is a solid that sublimes at 402 °C and must be handled as a solid. DMT melts at 141 °C and can be distilled. Therefore, DMT can be purified by distillation. This makes high purity PET possible. Also, DMT can be shipped, stored, and transferred as a liquid. It is much easier to pump a liquid than to transfer a solid. Solids take up more room (solid terephthalic acid density is much less than molten DMT) and therefore less can be shipped in a certain truck volume and larger storage tanks are needed. It is more difficult to transfer a solid from storage to a reactor.

The largest use for PET is for fibers. Much clothing and carpeting is made from PET fibers. Another large use for PET is plastic water bottles and plastic soda bottles. There are many other uses for PET; film is another major application. North American polyethylene terephthalate capacity is estimated to be 10.3 billion pounds [12].

Coca-Cola uses PET that is made from bio-based ethylene glycol as a basis for its PlantBottle®, which they use for a portion of their soda and water bottles. The ethylene glycol is made by dehydration of plant-derived ethanol to make ethylene, which is then converted to ethylene oxide and then ethylene glycol. There is some ongoing research to make bio-based terephthalic acid; another route is to substitute the terephthalic acid with furan dicarboxylic acid, which can be derived from sugars. Avantium is starting a 40 metric ton per year pilot plant to make furan dicarboxylic acid and is in the planning stage for a 50,000 metric ton plant [13].

The second largest volume polyester is polybutylene terephthalate, PBT, made from dimethyl terephthalate and 1,4-butanediol with a titanium alkoxide catalyst. Butanediol can be prepared from acetylene and formaldehyde followed by reduction of the triple bond.

Acetylene Formaldehyde 1,4-butanediol

DMT + 2x CH_3OH

Because the longer butanediol imparts a little less rigidity and more flexibility than ethylene glycol, PBT has a lower T_m and T_g than PET. The extra molecular motion due to the butanediol is also a reason why PBT tends to crystallize faster than PET; a big advantage in injection molding. For PBT, T_m is about 220 °C and T_g is about 40 °C. PET is about 30 °C higher for each value. Changes in the carboxylic acid portion also influence thermal properties. Going to the more rigid, polyethylene naphthalate increases the T_m by about 20 °C versus PET. The more flexible polyethylene adipate decreases the T_g and T_m, each by more than 100 °C versus PET.

PBT has many applications, most of them being injection molded articles. PBT is a crystalline polymer and has excellent solvent resistance. It can be suitable for applications requiring heat capabilities approaching its melting point of 220 °C. When blended with glass fibers, PBT can be injection molded to give articles with excellent stiffness. These blends are used for many electrical applications such as connectors. PBT is also blended with other polymers to impart solvent resistance to the final article. One example of this is automobile bumpers with the PBT imparting gasoline resistance. Common tradenames for PBT are Valox® (SABIC) and Ultradur® (BASF).

Both PET and PBT are made by a bulk melt process, typically in continuous fashion. Using PBT as an example, dimethyl terephthalate and an excess of butanediol are combined at a temperature above the melting point of DMT, typically around 150–160 °C in the presence of a tetraalkyltitanate. The reaction progress can be monitored by the distillation of methanol. As the reaction proceeds, the temperature is increased. Initially, the butanediol reacts, methanol is removed, and oligomeric esters are formed. Next, vacuum is applied and butanediol is removed by distillation. The molecular weight increases as oligomers combine in transesterification reactions. The reaction is allowed to proceed until the desired molecular weight is achieved, usually determined by monitoring the viscosity of the molten polyester. The molten polyester is then extruded through a die and pelletized.

8.6 POLYCARBONATE

Polycarbonate was invented in 1953 in the United States at GE Plastics and at about the same time in Germany at Bayer. GE marketed the polymer as Lexan® and Bayer as Makrolon®. In 2007, GE sold its plastics business to SABIC (Saudi Basic Industries Corporation), but the Lexan® name is still used. About 30 years after the original invention, Dow Chemical Co. began offering polycarbonate under the name Calibre®. Polycarbonate is an amorphous polymer and has excellent clarity, dimensional stability, and toughness. It is used for bullet-resistant windows and automotive headlamps.

Polycarbonate is made by the reaction of bisphenol-A (BPA) with phosgene or an equivalent. The reaction can be done in an interfacial process by adding base and phosgene to a stirring mixture of bisphenol-A in a water/organic solvent combination. The organic solvent dissolves the polycarbonate as it forms and is typically a chlorinated solvent such as methylene chloride or trichloroethane. After the reaction, the aqueous layer and the solvent are removed to give solid polycarbonate as powder or pellets, depending on the method of solvent removal. If sodium hydroxide is used as the base, then the byproduct is sodium chloride which can be converted back to sodium hydroxide and chlorine. The sodium hydroxide is used as the base to make more polycarbonate and the chlorine is used to make more phosgene.

HO–BPA–OH + Cl–C(=O)–Cl (Phosgene) —(H_2O/NaOH, solvent, catalyst)→ Polycarbonate (x) + 2x NaCl

BPA Phosgene Polycarbonate

2 NaCl ⟶ 2 NaOH + Cl_2

Cl_2 + CO ⟶ Cl–C(=O)–Cl

Phosgene

Alternatively polycarbonate can be made in a melt process using diphenyl carbonate in lieu of phosgene. Mainly due to some issues with inferior properties such as color and thermal aging, the melt process was not used commercially to any significant extent until the early 1990s when these problems were solved. A 1:1 molar ratio of BPA and diphenyl carbonate is combined and heated in the presence of a catalyst. The BPA dissolves in the molten diphenyl carbonate and phenol is generated by a transesterification reaction. Vacuum is applied and removal of phenol shifts the equilibrium to the polycarbonate. The molten polycarbonate is extruded through a die and pelletized.

BPA + Diphenyl carbonate $\xrightarrow{\text{Catalyst}}$ Polycarbonate + 2x Phenol

Diphenyl carbonate can be made from phenol and phosgene, but if it is made by a non-phosgene route, it can avoid the use of the toxic gases phosgene and chlorine. Diphenyl carbonate can be made from dimethyl carbonate. Dimethyl carbonate is available from the reaction of methanol with carbon monoxide [14] or with carbon dioxide [15].

$CH_3OH \xrightarrow{\text{CO or } CO_2}$ Dimethyl carbonate $\rightarrow$ Diphenyl carbonate

Another advantage of the melt process is that it avoids the use of chlorinated solvents. Trace levels of residual chlorinated solvents can be a concern for polycarbonate used in applications such as water bottles [16]. The solvent levels are very low in manufactured polycarbonate and become even lower during subsequent melt processing into articles; however use of the melt process avoids the issue entirely. Both processes have the potential for residual BPA and BPA is a degradation product of polycarbonate, especially degradation by hydrolysis. There is currently some concern about the health effects of BPA and the topic is the subject of much scientific debate and many studies [17].

8.7 NYLON

The reaction between an amine and a carboxylic acid can, with removal of water, yield an amide. It is this reaction that is the basis for nylon production. Nylon is a polyamide. Commonly, nylons are prepared from aliphatic diamines and aliphatic dicarboxylic acids. For nylon A,B, A is the number of carbons in the aliphatic diamine and B is the number of carbons (including the carbonyl carbons) in the dicarboxylic acid. Thus, nylon 4,6 is derived from 1,4-diaminobutane and adipic acid.

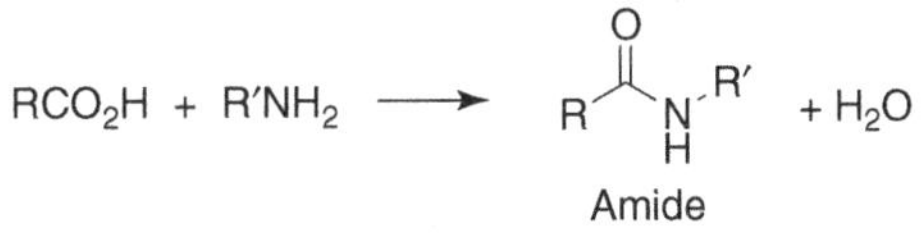

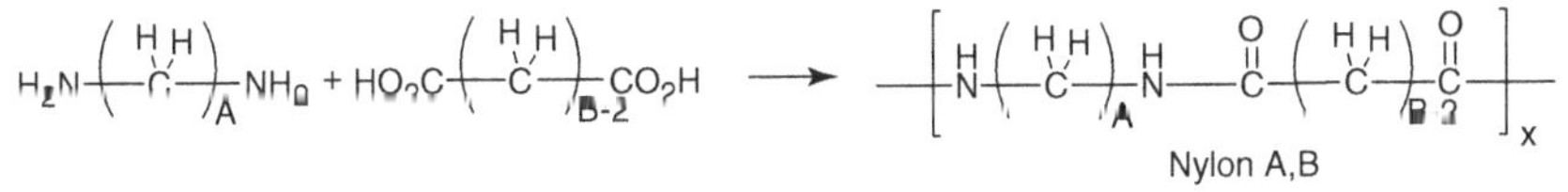

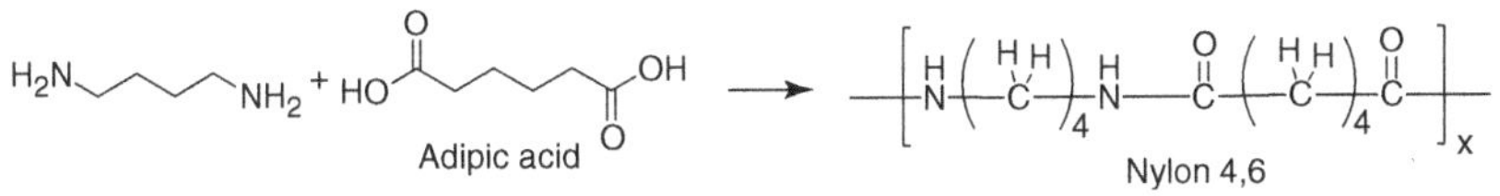

$$RCO_2H + R'NH_2 \longrightarrow RC(=O)NHR' + H_2O$$

Amide

$$H_2N{-}(CH_2)_A{-}NH_2 + HO_2C{-}(CH_2)_{B-2}{-}CO_2H \longrightarrow {-}[NH{-}(CH_2)_A{-}NH{-}C(=O){-}(CH_2)_{B-2}{-}C(=O)]_x{-}$$

Nylon A,B

$$H_2N(CH_2)_4NH_2 + HO_2C(CH_2)_4CO_2H \longrightarrow {-}[NH{-}(CH_2)_4{-}NH{-}C(=O){-}(CH_2)_4{-}C(=O)]_x{-}$$

Adipic acid — Nylon 4,6

The largest volume nylon is nylon 6,6. It has good properties and the starting materials are readily available. Adipic acid can be produced by the oxidation of cyclohexane. 1,6-Hexamethylene diamine is produced from adiponitrile. Adiponitrile can be produced from adipic acid or from butadiene [18].

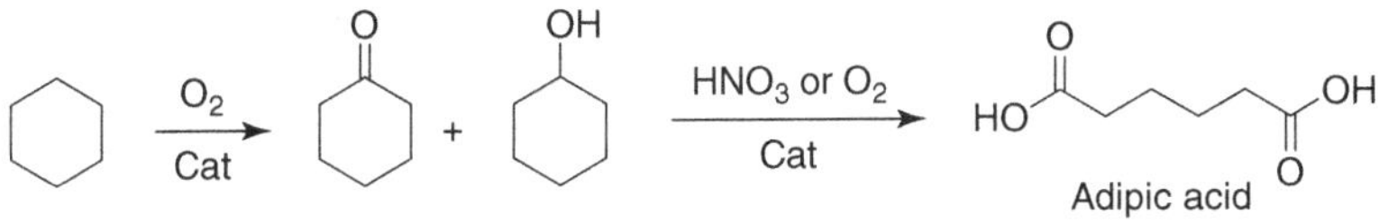

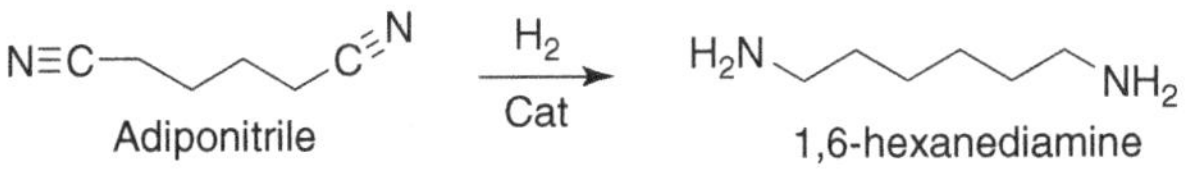

Cyclohexane $\xrightarrow[\text{Cat}]{O_2}$ cyclohexanone + cyclohexanol $\xrightarrow[\text{Cat}]{HNO_3 \text{ or } O_2}$ Adipic acid

$$N{\equiv}C{-}(CH_2)_4{-}C{\equiv}N \xrightarrow[\text{Cat}]{H_2} H_2N(CH_2)_6NH_2$$

Adiponitrile — 1,6-hexanediamine

The second largest volume nylon is nylon 6. Nylon 6 is not made from a diamine and a dicarboxylic acid, but from a six carbon cyclic amide. Cyclic amides are called lactams and caprolactam is a six-carbon lactam. Caprolactam can be made by the reaction of cyclohexanone with hydroxylamine, followed by a Beckmann rearrangement.

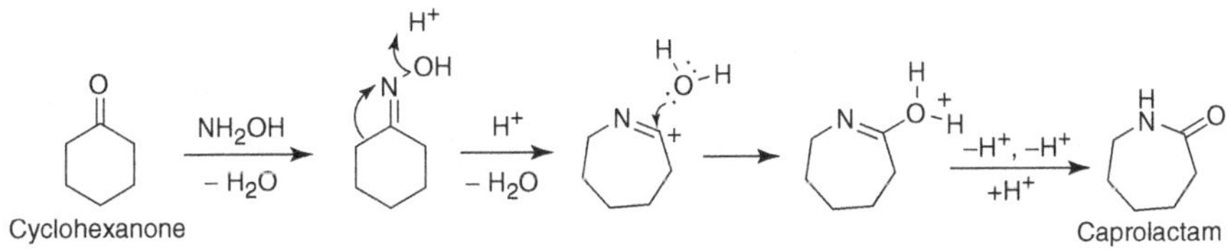

Cyclohexanone $\xrightarrow[-H_2O]{NH_2OH}$ oxime (H^+) $\xrightarrow[-H_2O]{H^+}$ nitrilium ion $\xrightarrow{H_2O}$ $\xrightarrow[+H^+]{-H^+,\,-H^+}$ Caprolactam

Caprolactam can be polymerized in the melt by use of a base and an activator. Two components, I and II, are used as starting materials. Component I is a catalyst-containing lactam melt and component II is an activator-containing lactam melt. The catalyst can be the sodium salt of caprolactam and the activator can be a derivative of caprolactam such as that prepared by the reaction of caprolactam with acetic anhydride or a diisocyanate [19]. The following example uses the derivative of caprolactam with hexamethylene diisocyanate.

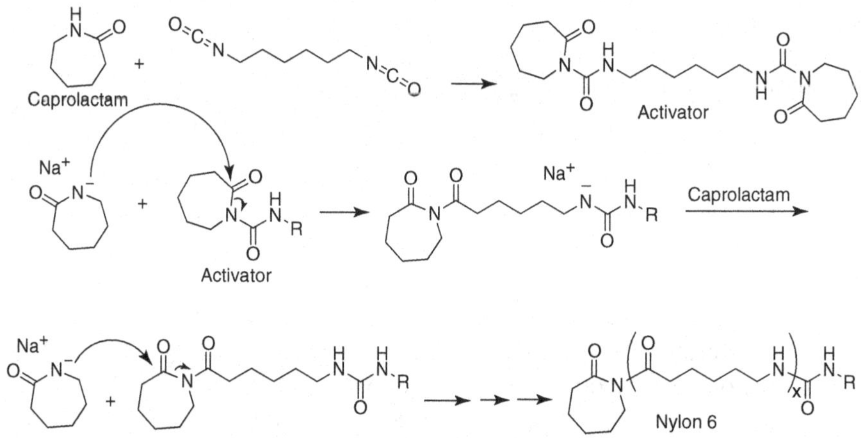

Caprolactam

Activator

Caprolactam

Activator

Nylon 6

Nylons have hydrogen bonding and this causes relatively high melting points for these crystalline polymers. The melting point of nylon 6,6 is 265 °C and nylon 6 is 225 °C [20]. Also, due to its structure, nylon absorbs much more water than many other plastics. For example, at 50% relative humidity, nylon 6,6 absorbs 2.8% (by weight) water and nylon 6, 3.0%. At saturation, nylon 6,6 absorbs 8.0% (by weight) water and nylon 6, 9.5%. Moisture absorption in nylons with fewer amide bonds per weight, such as nylon 6,12 or nylon 12, is lower. For example, nylon 12 absorbs 0.7% water at 50% relative humidity and 1.4% at saturation [21]. The water acts as a plasticizer, lowers the T_g, and affects properties. Long-chain nylons are produced at a much smaller scale and for specialized applications such as brake and gas lines that can bear the higher cost. For example, shipments of nylon 6 and 6,6 together are about 6 million metric tons annually but the combined market for nylon 11 and nylon 12 is more than an order of magnitude less at 100,000 metric tons [22]. Common nylon tradenames are Stanyl® (nylon 4,6; DSM), Novamide® (nylon 6, 6,6; DSM) and Zytel® (Dupont).

Aromatic polyamides are called aramids. These are produced in much lower volume than the aliphatic nylons, but because of their rigidity and high heat capabilities, are useful in certain specialty applications. Probably the two most common are the Dupont brands, Kevlar® and Nomex®. They can be made by treating an aromatic diamine with an aromatic diacid chloride [23, 24].

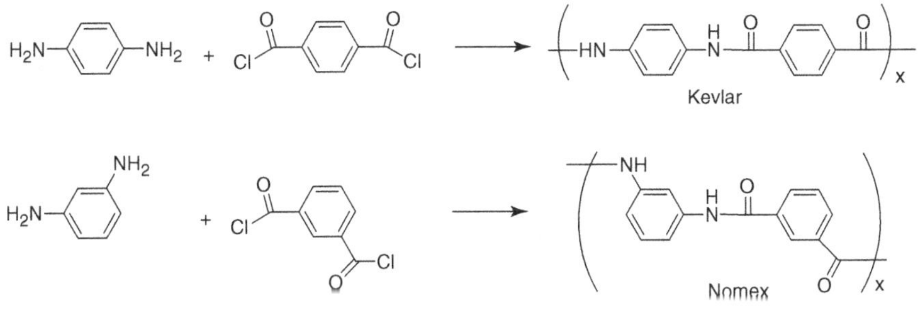

Kevlar

Nomex

Both can be spun into fibers. Kevlar is used in bullet resistant vests and Nomex is used in flame retardant protective clothing.

8.8 POLYIMIDE

An imide is made by reaction of an anhydride with an amine. In the first step, an amic acid is made. Upon heating, water is evolved and the amic acid closes to the imide. The reaction sequence is illustrated below with phthalic anhydride and aniline.

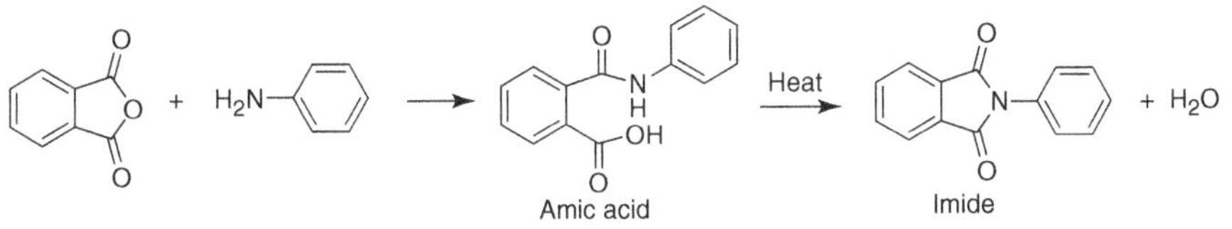

Amic acid

Imide

Polyimides are made similarly from diamines and dianhydrides. Dupont's Kapton® polyimide was the first commercially significant polyimide. It is made from pyromellitic dianhydride and oxydianiline. Kapton has good heat and strength properties and is used for specialty applications including space applications. It is used in electronics and in both film and tape applications.

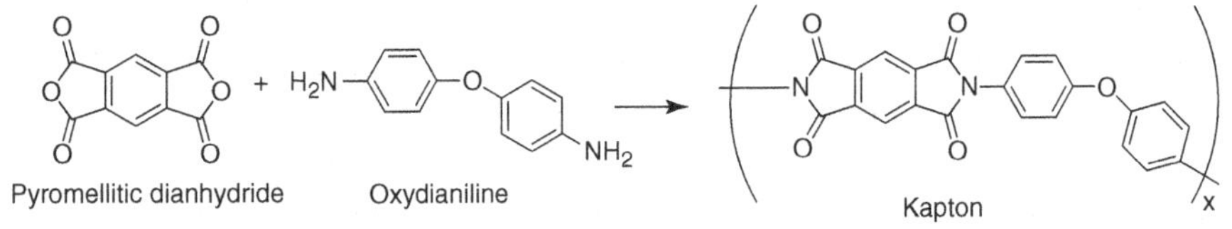

Pyromellitic dianhydride

Oxydianiline

Kapton

Another commercial polyimide, Ultem® was invented by GE scientists [25]. It is made from a derivative of BPA dianhydride and m-phenylene diamine. In 2010, SABIC (Saudi Basic Industries Corp., the purchaser of GE Plastics in 2007) opened a 5,000 metric ton per year Ultem® plant in Spain. SABIC also manufactures Ultem in Mt. Vernon IN and this new facility [26] increases their capacity by 35%.

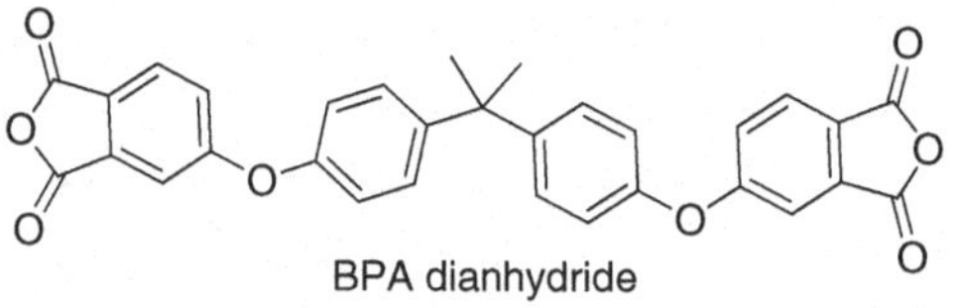

BPA dianhydride

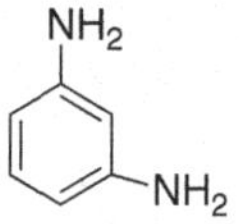

m-phenylene diamine

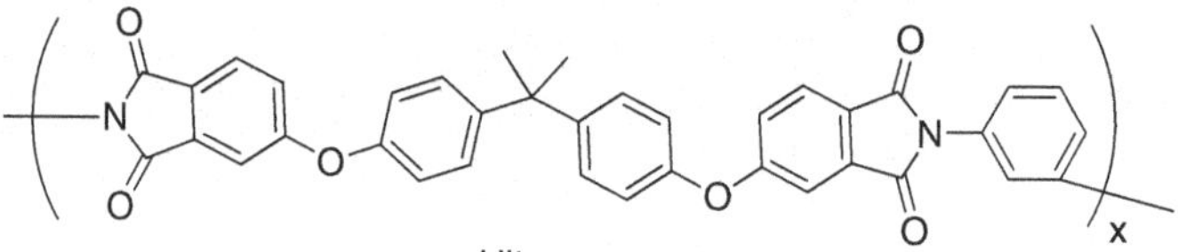

Ultem

For specific applications, various specialty polyimides have been prepared. For example, polyimidesiloxanes are useful for screen printing a paste for microelectronic applications [27].

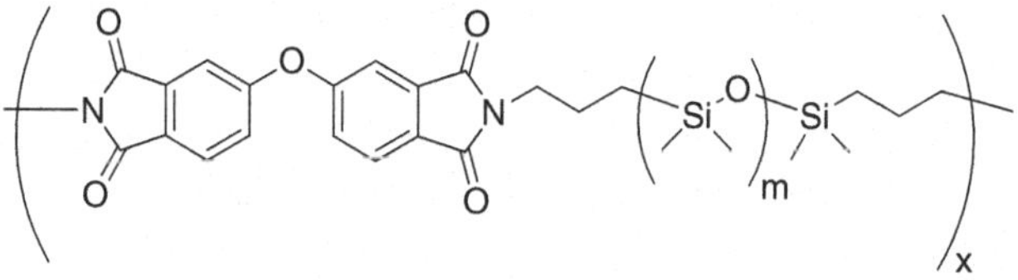

8.9 FLUOROPOLYMERS

Fluoropolymers have excellent heat, chemical and corrosion resistance. The most common is polytetrafluoroethylene (PTFE), often known by the Dupont trademark, Teflon®. Other tradenames include Dyneon® (3M) and Fluon® (Asahi Glass). The invention of PTFE is often used as an example of serendipity, but it was actually a combination of serendipity, curiosity, and hard work. Roy Plunkett was working on experimental refrigerants when a cylinder that had been filled with tetrafluoroethylene (TFE) gas did not deliver gas when the valve was opened. Often, when something does not occur as planned, people discard the results and move on, but Plunkett was curious. When the cylinder was cut open, a white lubricious solid was discovered. Further investigation revealed the solid to be a polymer of tetrafluoroethylene [28].

Consumers know PTFE because it is used as a coating material in non-stick cookware. The excellent chemical resistance makes it suitable for coating reactors and piping that are exposed to aggressive chemicals. The extremely low coefficient of friction makes it useful for bearings and gears. Specialty bullets are coated with PTFE to reduce wear on a rifle barrel.

PTFE is available either as the homopolymer or modified with small amounts (typically less than 1% by weight) of a comonomer. The comonomer can have the benefit of improving processability by decreasing the molecular weight. Crystallization is also inhibited. The modifier is used at a low enough level that the desirable PTFE properties are maintained.

$$CF_2{=}CF_2 + CF_2{=}CFR \longrightarrow \text{PTFE}$$

TFE Optional modifying modifier

$R = CF_3, OCF_2CF_2CF_3, OCF_2CF_3$, etc.

As with PVC, PTFE is made by a radical polymerization and can be made by a suspension process or an emulsion process. The emulsion polymerization gives an aqueous dispersion of PTFE and is sometimes referred to as a dispersion polymerization. The TFE is dispersed in an agitated aqueous medium, using a surfactant and a water soluble free radical initiator. Often a hydrocarbon such as molten paraffin wax is present to retard coagulation of the PTFE [29]. Typically, the surfactant has been a perfluoroalkanoic acid or salt, commonly ammonium salts, such as the ammonium salt of perfluorooctanoic acid (PFOA). PFOA has also been used in the textile industry to impart water and oil repellency. PFOA is persistent in the environment and has been found in blood samples of the general United States population so the industry has been working toward elimination of plant emissions and product content. Other surfactants, many of them based upon ethers have been described. For example, $CF_3O(CF_2)_3OCHFCF_2CO_2^- NH_4^+$ has been described as a TFE polymerization surfactant that in rat studies had a much lower tendency to bioaccumulate than the ammonium salt of PFOA [30]. PTFE can also be prepared by a suspension process [31]. This usually involves vigorous agitation of an aqueous medium and yields granulated particles.

The carbon–fluorine bond is much stronger than the carbon–hydrogen bond (485 kJ/mol for C–F; 413 kJ/mol for C–H). Because of this, the C–F bond is much less likely to be broken by homolytic cleavage. This is a major contributor to the excellent UV, thermal, and oxidative stability of PTFE. It also means that in the radical polymerization, chain termination reactions are less likely. Therefore, PTFE tends to polymerize to high molecular weights which in turn causes high viscosity and problems with melt processing. To make lower molecular weight polymers, molecular weight regulators, also known as chain transfer agents, can be added to the polymerization. Examples of chain transfer agents include hydrogen and compounds with a C–H bond such as alkanes.

The PTFE chain adopts a slowly twisting helix with an outer sheath of fluorine atoms encompassing a carbon-based core. There is mutual repulsion of the fluorine atoms. These cylinders can slip past each other and this can lead to cold flow or creep [32]. Creep is permanent deformation of a plastic due to the prolonged application of stress. An example is a plastic strap that is holding a weight and over time permanently stretches and is elongated even after the weight is removed. PTFE is more susceptible to creep than many other polymers.

Porous PTFE, also known as expanded PTFE (ePTFE) can be highly porous while retaining excellent strength. Fluorocarbons are hydrophobic so the combination of porosity and hydrophobicity makes ePTFE excellent for waterproof clothing. This technology is the basis for the outdoor clothing made from Gore-Tex® [33–35].

There are several commercial copolymers of TFE such as the copolymer of TFE with hexafluoropropylene (HFP), referred to as fluorinated ethylene propylene (FEP). FEP has lower melt viscosity, enabling conventional melt processing and is often used for wire insulation and molded parts. There is a tradeoff with lower thermal stability and lower upper use temperature. Another copolymer is TFE with perfluoro(propyl vinyl ether). This is commonly referred to as a perfluoroalkoxy (PFA) polymer. PFA polymers are melt-processable and generally have better thermal stability and high temperature properties than FEP polymers. PFA is commonly used as pipe liners, tubing, and films. Copolymer side chains such as the perfluoropropoxygroup and trifluoromethyl group increase the resistance of polymer chains sliding past each other and therefore improve the creep resistance.

Other fluorinated polymers include polyvinylidenedifluoride (PVDF), made by the polymerization of 1,1-difluorethylene. PVDF is known by its tradenames, Kynar® (Arkema) and Hylar® (Solvay) and is less expensive. It has excellent chemical resistance and is used as a coating for metals and for wire jacketing.

8.10 POLYPHENYLENE SULFIDE

There have been several earlier attempts to make polyphenylene sulfide (PPS), but in the late 1960s, workers at Phillips Petroleum developed the first commercial process [36]. Sodium sulfide reacts with p-dichlorobenzene in a polar organic solvent. Common trade names are Ryton® (Phillips Petroleum) and Fortron® (Fortron Industries, a joint venture of Hoechst Celanese and Kureha Chemical).

$$Cl{-}C_6H_4{-}Cl + Na_2S \longrightarrow \left[{-}C_6H_4{-}S{-} \right]_n$$

PPS

PPS is particularly well suited to demanding applications where it must stand up to high temperatures and solvent attack while maintaining overall mechanical and dimensional integrity. PPS has continuous use temperatures exceeding most other thermoplastics. It has a low coefficient of thermal expansion, good inherent flame retardance and good electric properties. One drawback of PPS is that it is relatively brittle. It is usually compounded with glass or carbon fibers or with mineral fillers to improve its impact strength and other mechanical properties. Because of its thermal stability and resistance to automotive fluids, filled PPS has many applications in automotive parts such as switches, alternators and fuel pump housings.

8.11 ACETAL RESIN

Acetal resin has a good combination of strength and stiffness, good creep resistance, good fatigue resistance, abrasion resistance, and low wear and friction. Creep is the tendency of a material to deform under load. For example, a plastic grocery bag made of polyethylene may not stretch if it has a heavy object in it. However, if suspended for a period of time it may permanently stretch and deform. This is more likely to happen as temperature increases. Creep is deformation as a function of time for a sample under constant stress. Fatigue is the progressive damage that occurs when a material is subjected to cyclic loading. Consider a piece of uncoated copper wire. If you bend it, it doesn't break. However, if you repeatedly bend it back and forth, it will eventually break. This is an example of fatigue. Fatigue resistance is important for a material that is repeatedly flexed back and forth. Creep resistance is important for a material, such as a hanging strap, that is constantly subjected to a load. Because of its properties, acetal resin is often used for moving parts such as gears or zippers.

Acetal resin is either a homopolymer or copolymer of formaldehyde. The homopolymer is sometimes called polyoxymethylene or POM. It is made by polymerizing either anhydrous formaldehyde or its cyclic trimer trioxane. The polymerization can be done using either acidic or basic catalysts. At least some of the end groups are hydroxyl groups and with heat, the resin can depolymerize back to formaldehyde. This was an early problem, essentially making the polymer useless. The thermal stability problem was solved by a post treatment where the hydroxyl ends were capped, for example by treatment with acetic anhydride [37].

$$H_2C{=}O \text{ or } (CH_2O)_3 \longrightarrow \left(\begin{matrix} C - O \\ H\;H \end{matrix} \right)_x$$

$$HO\left(\begin{matrix} C - O \\ H\;H \end{matrix} \right)_x H \xrightarrow{Ac_2O} H_3C-\overset{O}{\overset{\|}{C}}-O\left(\begin{matrix} C - O \\ H\;H \end{matrix} \right)_x \overset{O}{\overset{\|}{C}}-CH_3$$

Acetal resins do have some weaknesses. The acetal linkage can be hydrolyzed, especially under acidic conditions, back to formaldehyde. The methylene is prone to oxidation especially if chlorine is present. In the 1970s and 1980s, acetal resins were used for plumbing fittings. Due to hydrolysis, and attack by the chlorinated water, these parts began to fail resulting in severe damage and class action law suits. Their use as fittings was discontinued, but more recently there have been acetal failures in toilet valve assemblies.

8.12 THERMOSETS

The polymers discussed to this point are reasonably linear. They can be remelted or redissolved and are members of the general class of thermoplastic polymers. For example, a used PET water bottle can be chopped into flakes and then remelted and fabricated into another article such as a flower pot, golf tees, or many other items. These thermoplastics are predominantly linear. Even in the case of polyethylene, where there is often some branching, they soften when heated and can be reshaped or molded into new articles. However, if the polymer is not linear but three dimensional it cannot be reshaped. These polymers are formed in the mold and take a set. After the polymerization is complete, they cannot be remelted and reshaped. These are called thermoset polymers or thermoset resins.

One type of thermoset polymer is polyurethane. Polyurethane can be linear or three dimensional, depending on the starting materials for the polymerization. The urethane linkage is also known as a carbamate. The polymers are always called polyurethanes and not polycarbamates. A carbamate can be prepared by the reaction of an alcohol with an isocyanate. Isocyanates are readily prepared from an amine and phosgene. The reaction sequence is illustrated with aniline which forms the isocyanate and then reacts with ethanol to form the carbamate or urethane linkage. A competing reaction is the trimerization of isocyanates. The reaction between an isocyanate and an alcohol is a rapid reaction, but it is typically catalyzed, often with a tertiary amine. The catalyst is often chosen to give the desired amount of trimerization.

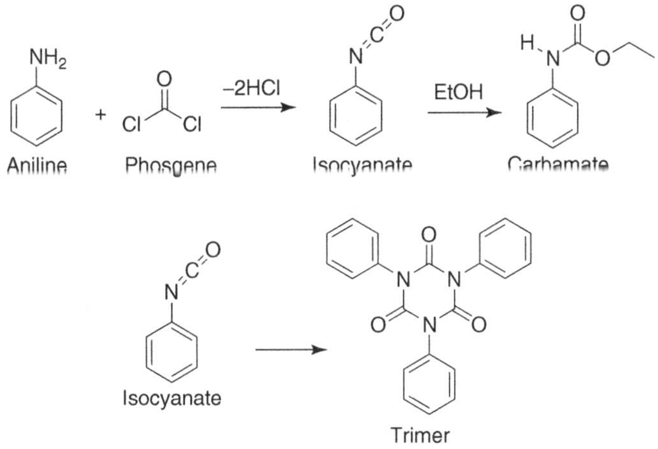

Aniline + Phosgene —($-2HCl$)→ Isocyanate —(EtOH)→ Carbamate

Isocyanate → Trimer

Diisocyanates are commonly used with either diols or a mixture of diols and polyols. When diols are exclusively used and the trimerization reaction is minimized, the polyurethane is substantially linear, but it becomes increasingly three-dimensional with increasing levels of polyol or with trimerization of the diisocyanate. The trimerization reaction and the use of polyols enable thermoset polyurethanes. Common diisocyanates include toluene diisocyanate (TDI), methylene diphenyl diisocyanate (MDI), naphthalene diisocyanate (NDI), and hexamethylene diisocyanate (HDI). TDI is prepared by the nitration of toluene to give a mixture of mainly 2,4- and 2,6-dinitrotoluene, followed by reduction and reaction with phosgene. The isomers are typically not separated and are used as a mixture.

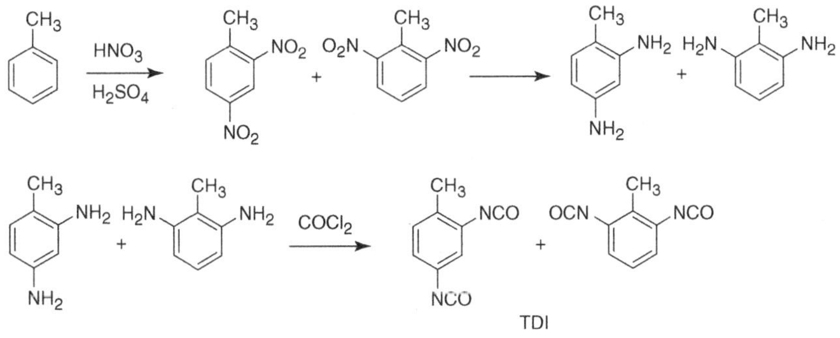

TDI

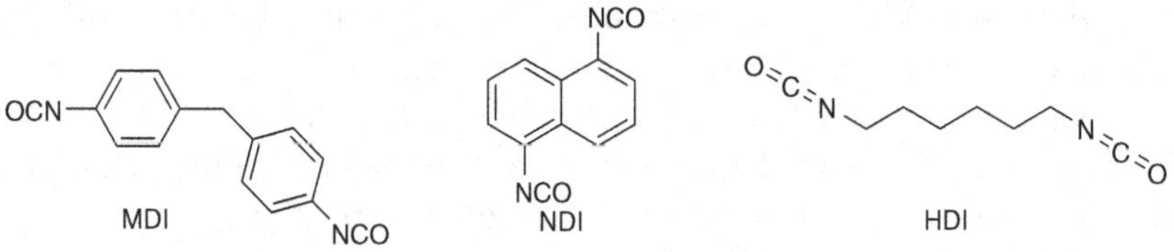

MDI NDI HDI

When a diisocyanate trimerizes, one isocyanate from each of the monomers forms the trimeric ring and the other remains available for polymerization. The trimer effectively contains three isocyanates and therefore allows the polymer to grow in three directions. This is illustrated below with the trimer of HDI.

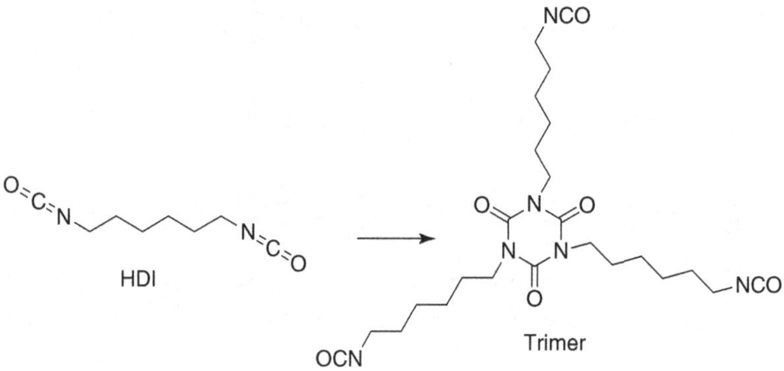

HDI → Trimer

Depending on the desired properties, the diol can be a short chain diol such as ethylene glycol, 1,4-butanediol, or 1,6-hexanediol or a hydroxyl terminated oligomer such as a polyether diol or a polyester diol. When a polyol is used, similarly it can be a short chain polyol such as glycerol or it can be an oligomeric polyol such as polyether polyols or polyester polyols.

Both the diisocyanate portion and the diol/polyol portion are liquids. One common way to fabricate polyurethanes is by use of reaction injection molding (RIM). With this method, each of the two liquid components is passed through a mixer and into a mold where the reaction takes place to form the polyurethane.

Most polyurethane applications are based upon foamed polyurethanes. Examples include mattresses, seat cushions, and insulation. Although in principle any volatile material that can create bubbles can be used to foam a polymer, carbon dioxide is commonly used because it can be easily generated by the reaction of the isocyanate linkage with water. The byproduct is an amine which can react with an isocyanate to form a urea, thereby continuing the polymer growth. The chemistry is illustrated below.

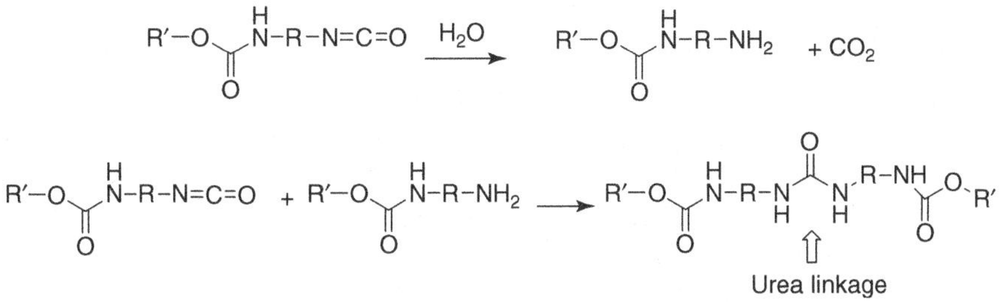

$$R'{-}O{-}C(=O){-}NH{-}R{-}N{=}C{=}O \xrightarrow{H_2O} R'{-}O{-}C(=O){-}NH{-}R{-}NH_2 + CO_2$$

$$R'{-}O{-}C(=O){-}NH{-}R{-}N{=}C{=}O + R'{-}O{-}C(=O){-}NH{-}R{-}NH_2 \longrightarrow R'{-}O{-}C(=O){-}NH{-}R{-}NH{-}C(=O){-}NH{-}R{-}NH{-}C(=O){-}O{-}R'$$

Urea linkage

Another type of thermoset is the phenol formaldehyde resins, also known as phenolic resins. They are the first truly synthetic polymer [38] to be developed and are often referred to as Bakelite after Leo Bakeland, who invented them in 1907. Phenolic resins are made by reacting phenol with formaldehyde under acidic or basic conditions. Under basic conditions, a greater than a 1:1 molar ratio of formaldehyde to phenol is used and the product is called a resole. Under acidic conditions, less than a 1:1 molar ratio of formaldehyde to phenol is used and the product is called a novolak [39]. Under basic conditions, the reaction proceeds by an aldol condensation type mechanism to give the hydroxymethylphenol. This in turn can react with another phenol to give a methylene bridged bisphenol and then with more formaldehyde and phenol to give the resole.

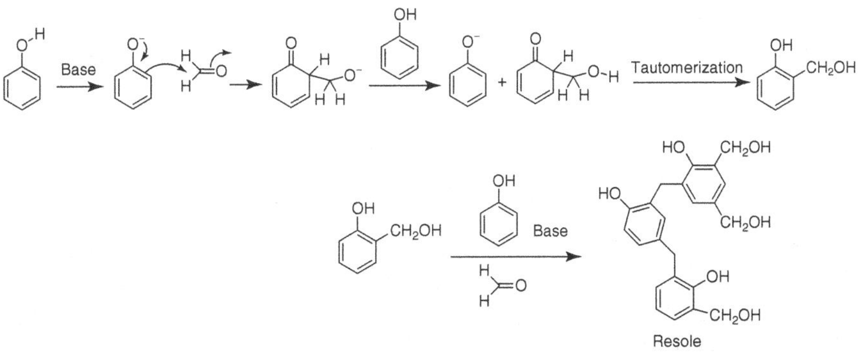

Base

Tautomerization

CH_2OH

Base

Resole

The resole is not specifically the compound shown, but rather a mixture of that and similar structures. The resole can then be heated in a mold to give further condensations and a thermoset polymer.

Under acidic conditions, phenol reacts with formaldehyde by an electrophilic aromatic substitution mechanism. This in turn can react with another phenol to give a methylene bridged bisphenol and then with more formaldehyde and phenol to give the novolak. The novolak is not specifically the compound shown, but rather a mixture of that and similar structures.

The novolak can then be treated with crosslinkers such as hexamethylenetetramine in a mold to give further condensations and a thermoset polymer. With both the resole and the novolak the structures can be affected by the reaction conditions and the choice of catalyst. In particular, the ortho:para ratios can vary. Resoles are characterized by a large number of methylol groups which can condense with the phenolic rings to evolve water and give additional methylene bridges in the cure step.

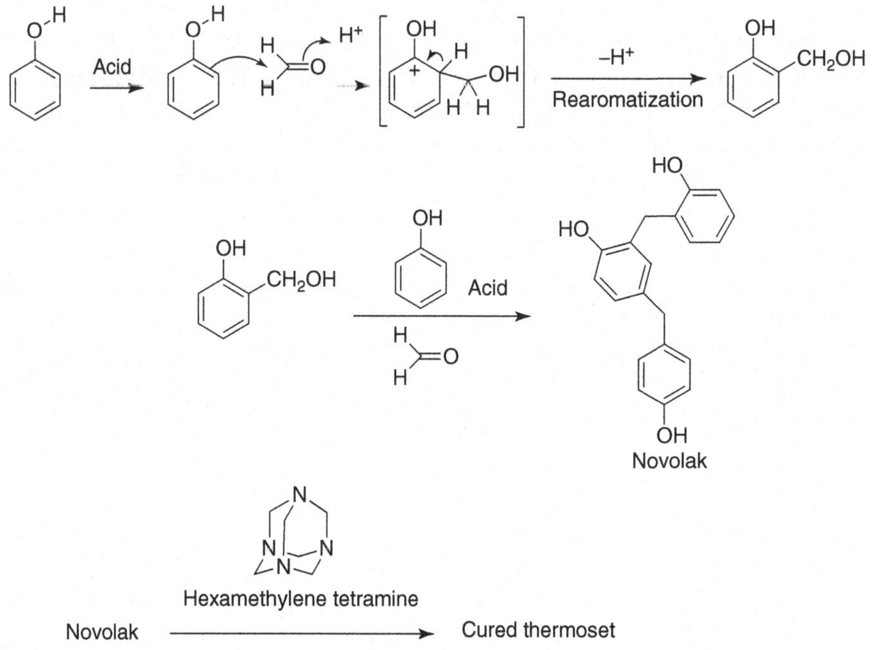

Acid H^+ $-H^+$ Rearomatization CH_2OH

Acid

Novolak

Hexamethylene tetramine

Novolak → Cured thermoset

Phenolic resins are useful as adhesives, especially for wood composites such as plywood or particle board. They have high heat capabilities and excellent flame resistance; this makes them useful for brake linings and clutch facings and more mundane applications such as oven knobs and pot handles. There are many other applications including printed circuit boards and as an abrasive such as in waterproof sandpaper.

Epoxy resins are thermoset polymers prepared by the reaction of an epoxy compound, typically a low molecular weight diepoxide, with a curing agent. These two parts are combined and a chemical reaction causes the cure. They are often called two-part epoxies. They come in two tubes for home use. There are many variants of epoxy compounds and curing agents but diglycidyl ethers of bisphenols cured with diamines are common. The diglycidyl ethers are readily prepared from epichlorohydrin. The reaction is illustrated with bisphenol A diglycidyl ether as the epoxy compound and 4,4′-diaminodiphenylsulfone as the curing agent.

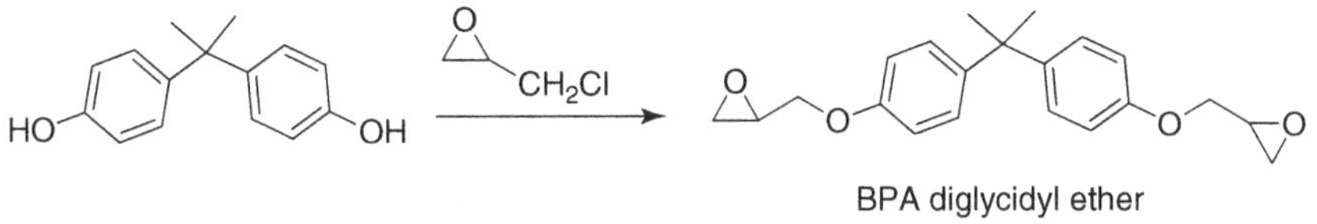

The amine reacts with the epoxide by opening the epoxide to an alcohol.

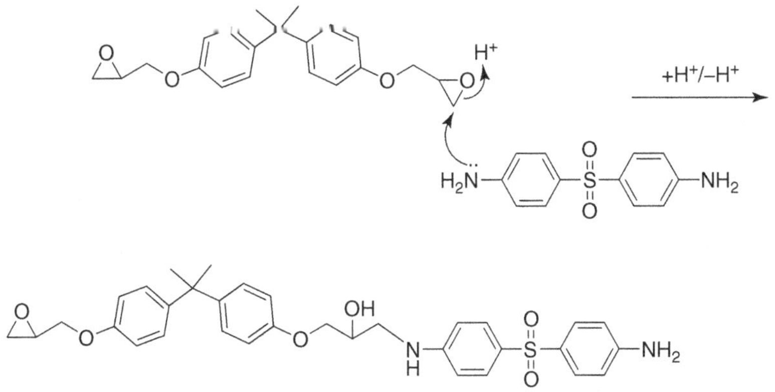

This reaction continues with each of the primary amines reacting with two epoxides. It is shown with 4,4′-diaminodiphenylsulfone reacting with four equivalents of bisphenol A diglycidyl ether, but recognize that each of the epoxide ends can react.

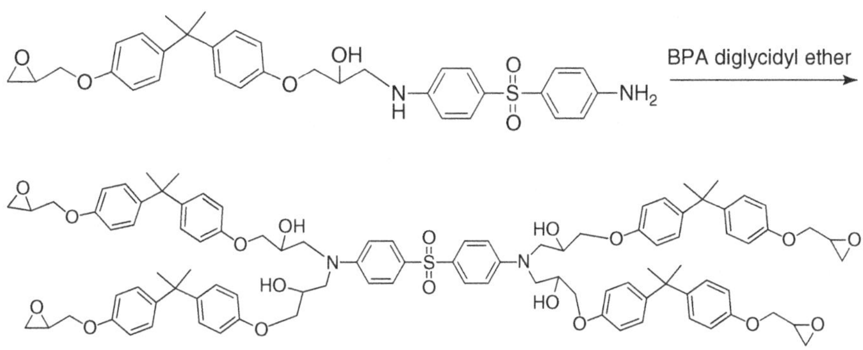

The net result is that the polymer extends in many directions forming a three-dimensional crosslinked network of the thermoset polymer. Note that, in contrast to the phenolic thermosets, no water is evolved in the cure step.

This is an advantage for many applications, and epoxides are widely used in microelectronic applications. The structure of the cured resin is highly polar and can form hydrogen bonds. This makes the cured resin an excellent adhesive.

QUESTIONS

1. Which of the following polymers do **not** depend upon ethylene (or a derivative of ethylene) as a raw material (select all that apply; could be any number from 0 to 5 choices)?
   a. polystyrene
   b. PVC
   c. polyethylene terephthalate
   d. nylon 4,6
   e. polybutylene terephthalate
2. Dimethyl terephthalate has two major advantages over terephthalic acid. What are they?
3. Draw the starting monomer or monomers to make nylon 4,6.
4. Write the structure for polycarbonate
5. Which of the following polymers are addition polymers (select all that apply; could be any number from 0 to 5 choices)?
   a. polystyrene
   b. PVC
   c. polyethylene terephthalate
   d. nylon 4,6
   e. polybutylene terephthalate
6. If dimethylterephthalate costs 45.0¢ per pound, 1,4-butanediol costs 60.0¢ per pound, and the price of the catalyst is negligible, assuming 100% yield, calculate the raw material cost of polybutylene terephthalate to the nearest tenth of a cent.
7. Draw the starting monomer or monomers to make nylon 6,10.
8. Start with chemicals directly obtained from petroleum and show the sequence of steps (don't need to show mechanism; but show major isolated intermediate chemicals) to make PET.
9. Assume that the amount of base and N-acetyl caprolactam used is so small that it is insignificant in the weight of the polymer and in the cost.

Assuming 100% yield, if the cost of caprolactam is 68.0¢ per pound, what is the raw material cost of nylon 6 (report your answer to three significant figures)?

10. Classify each of the following as "A" for addition polymer or "C" for condensation polymer.

polyethylene

nylon 6,10

polycarbonate

polybutylene terephthalate

11. Which is the correct name for the polymer made from adipic acid and diaminobutane?

a. nylon 4

b. nylon 4,4

c. nylon 4,6

d. nylon 6,4

e. nylon 6,6

f. nylon 6

12. Starting only with materials isolated directly from petroleum or from mining (can use air, water, radical initiators, and acid without showing the source), sketch the reactions to show the synthesis of polycarbonate. When you write the final polymer structure, make sure you use the parentheses correctly.

13. Write the structure, making sure you use the parentheses correctly, of a polyimide.

14. Polyurethanes are often made by reaction injection molding of two components. One component is the alcohol component and is typically a mixture of diols with triols or polyols. What is the other major component? Give both the name and the generalized structure.

15. This is made by a high temperature high pressure radical polymerization of ethylene.

a. HDPE

b. LDPE

c. LLDPE

d. EEA

16. This is made by using an organometallic catalyst and polymerizing ethylene with enough 1-hexene to adjust the density to 0.92.

a. HDPE
b. LDPE
c. LLDPE
d. EVA

17. (6 points) Of the four listed, this polyolefin has the best adhesion.
a. HDPE
b. LDPE
c. LLDPE
d. EEA

18. This major polymer is made by an emulsion or suspension radical process, and generally has a heat stabilizer added. The polymer has less than half of its mass from petroleum feedstocks. What is the polymer?

19. ABS terpolymer has good heat stability, toughness, and stiffness. ABS is an acronym for which three monomers?

20. Thermoset epoxies are commonly made by mixing a bis-epoxy component with:
a. a bisphenol
b. a diisocyanate
c. a diamine
d. formaldehyde

21. Toluene 2,4-diisocyanate is formed from the reaction of 2,4-diaminotoluene with phosgene. Write the structure for toluene 2,4-diisocyanate. What is it used for?

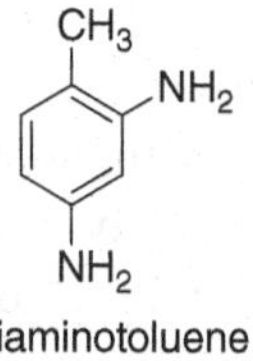

CH_3 NH_2 NH_2

2,4-diaminotoluene

22. Between nylon 6,6 and PTFE which polymer absorbs more water? Why?

23. What do the letters LLDPE stand for?

24. How is LLDPE made?
a. Ziegler–Natta process
b. radical process
c. $TiCl_4$ catalyst
d. with a vinyl acetate comonomer

25. Polyethylene manufacturers that want to lower the density often copolymerize ethylene with what other hydrocarbon (hint: formula = $C_nH_{2n)}$?

REFERENCES

1. PC Painter, MC Coleman. *Fundamentals of Polymer Science*. 2nd ed. CRC Press, Boca Raton, FL; 1997:45.
2. S Nagy, J Tyrell. U.S. Pat. No. 6,462,153. 2002.
3. L Cohen. U.S. Pat. No. 4,256,864. 1981.
4. PC Painter, MC Coleman, *Fundamentals of Polymer Science*. 2nd ed. CRC Press, Boca Raton, FL; 1997:9.
5. WH Starnes. Journal of Polymer Science: Part A: Polymer Chemistry 2005; 43:2451–2467.
6. C Carraher. Seymour/Carraher's Polymer Chemistry. 7th ed. CRC Press, Boca Raton, FL; 2008:17.
7. P Morse. Chemical and Engineering News 2011; 89(22):28.
8. Agency for Toxic Substances and Disease Registry (ATSDR). Toxicological Profile for Di(2-ethylhexyl)phthalate (DEHP). Atlanta, GA: U.S. Department of Health and Human Services, Public Health Service; 2002.
9. Carl Dupre. U.S. Pat. No. 4,146,589. 1979.
10. Jae Goo Doh.U.S. Pat. No. 7,393,889. 2008.
11. Thomas Huver, Christian Nicolaisen, Susanne Camp. U.S. Pat. No. 5,561,198. 1995.
12. Alexander Tullo. Chemical and Engineering News 2011; 89(6):19.
13. Alexander Tullo. Chemical and Engineering News 2012; 90(4):20.
14. Nicola Di Muzio, Carlo Fusi, Franco Rivetti, Giacomo Sasselli. U.S. Pat. No. 5,210,269. 1993.
15. Kazuto Kobayashi, Hiroyuki Osora, Yoshio Seiki, Masaki Iijima. U.S. Pat. No. 7,605,285. 2009.
16. TJ Stanley, MM Alger.Ind. Eng. Chem. Res.1989; 28:865–869.
17. U.S. Food and Drug Administration Update on Bisphenol A (BPA) for Use in Food: January 2010.
18. K Weissermel, H-J. Arpe. *Industrial Organic Chemistry*. 3rded. New York: VCH Publishers Inc.; 1997:240–241, 244–249.
19. P Horn, H Bunsch, R Gehm, M Marx. U.S. Pat. No. 4,598,125. 1986.
20. C Carraher. Seymour/Carraher's Polymer Chemistry. 7th ed. CRC Press, Boca Raton, FL; 2008:17.
21. Harold F Giles Jr., Eldridge M Haber, John R Wagner Jr. Extrusion: *The Definitive Processing Guide and Handbook*. (Plastics Design Library). Pub: William Andrews, Inc., Norwich, NY; 2007:209.

22. Alexander Tullo. Chem. and Eng. News 2013; 91(7):28.
23. SL Kwolek, PW Morgan, WR Sorenson. U.S. Pat. No. 3,063,966. 1962.
24. HB Hockessin. U.S. Pat. No. 3,767,756. 1973.
25. P. Gallagher. U.S. Pat. No. 4,550,156. 1985 and references therein.
26. Alexander Tullo. Chem. and Eng. News 2010; 88(40):22.
27. J Choi, J Tyrell, P Dubell. U.S. Pat. No. 5,554684. 1996.
28. RJ Plunkett. U.S. Pat. No. 2,230,654. 1941.
29. CW Jones. U.S. Pat. No. 6,136,933. 2000.
30. K Hintzer, et al. U.S. Pat. No 8,119,750. 2012.
31. RM Aten. U.S. Pat. No. 5,405,923. 1995.
32. John Scheirs. Modern Fluoropolymers. New York: Wiley; 1997:2–3.
33. RW Gore. U.S. Pat. No. 3,953,566. 1976.
34. RW Gore. U.S. Pat. No. 4,187,390. 1980.
35. RW Gore, SB Allen. U.S. Pat. No. 4,194,041. 1980.
36. James T Edmonds, Harold W Hill. U.S. Pat. No. 3,354,129. 1967.
37. Stephen Dal Nogare, John Punderson. U.S. Pat. No. 2,998,409. 1961.
38. Louis Pilato. *Phenolic Resins: A Century of Progress*. Springer, Heidelberg, Germany; 2010:1.
39. Louis Pilato. *Phenolic Resins: A Century of Progress*. Springer, Heidelberg, Germany; 2010:41.

CHAPTER 9

Blends and Additives

9.1 BLENDS

Polymers are often blended together to obtain the desirable properties of each component. It is more expedient to blend commercial polymers together than to produce a new polymer. There are many commercial success stories based upon this strategy, but it isn't necessarily true that properties are optimized. Consider a handsome albeit lazy man married to a highly intelligent but unattractive woman. They could have a beautiful intelligent baby if the baby has the father's looks and the mother's brains. They could also have a baby that is lazy and unattractive if the baby inherits the father's work ethic and the mother's looks. So it can be with polymer blends. Much research is needed to optimize a polymer blend. If one polymer is completely insoluble in the other and there is no adhesion between the polymer domains, an article molded from the blend can delaminate. For a good blend, one polymer is either soluble in the other, a miscible blend, or if insoluble, an immiscible blend, there needs to be good adhesion and dispersion of the polymer domains.

Polyphenylene oxide (PPO) is a high heat polyether made by oxidatively coupling 2,6-xylenol. The polymer was invented by Allan Hay of General Electric [1–3] and is the basis for the Noryl® product line.

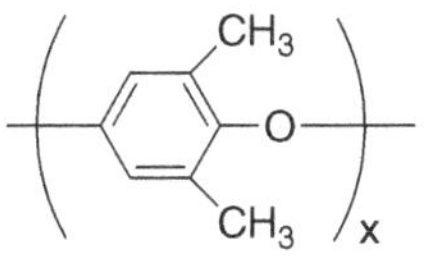

PPO

PPO [4] is a high heat polymer and has a Tg of 208 °C. However, it is a brittle polymer and is very difficult to process. By itself, it has limited commercial value. However, PPO is miscible with polystyrene and addition of

Fundamentals of Industrial Chemistry: Pharmaceuticals, Polymers, and Business, First Edition. John A. Tyrell.

polystyrene lowers the Tg. When two polymers are miscible, they have a single Tg which is intermediate between the Tg of each of the component polymers. The greater the amount of polystyrene added, the lower the Tg of the PPO blend. Rather than using polystyrene itself, Noryl uses high impact polystyrene (HIPS). HIPS is prepared by polymerizing styrene in the presence of polybutadiene, thereby forming a graft copolymer in with polybutadiene and polystyrene homopolymer. There are different grades of Noryl, but a Tg of about 150 °C is typical for this miscible blend.

Another blend, also invented at General Electric, is Xenoy®. This is an immiscible blend of polybutylene terephthalate (PBT) and polycarbonate (PC) [5, 6]. This blend is immiscible; it has separate phases of PBT and of PC. An immiscible blend is characterized by two distinct Tgs and thermal analysis of this blend shows the Tg of PBT and of PC as well as a sharp melting point for the crystalline PBT. The blend also includes an impact modifier. It exhibits the good impact associated with PC and the solvent resistance associated with PBT. Because of these properties, it is suitable for applications such as automotive bumpers requiring high impact and gasoline resistance.

The carbonate linkage can react with the ester linkage in a transesterification reaction. This reaction is illustrated by the reaction of diphenyl carbonate with butylene dibenzoate to form phenyl benzoate.

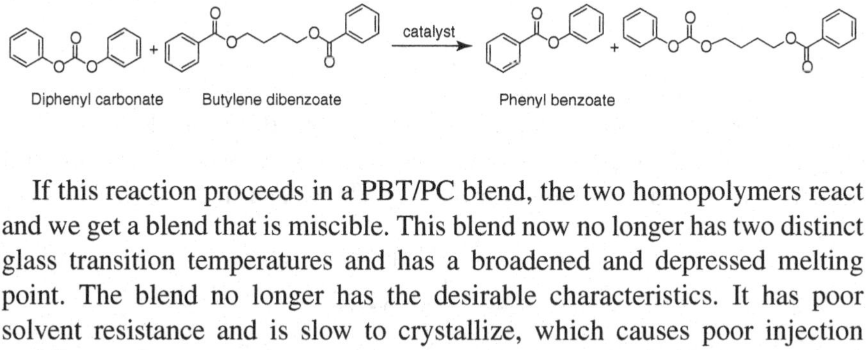

catalyst

Diphenyl carbonate + Butylene dibenzoate → Phenyl benzoate +

If this reaction proceeds in a PBT/PC blend, the two homopolymers react and we get a blend that is miscible. This blend now no longer has two distinct glass transition temperatures and has a broadened and depressed melting point. The blend no longer has the desirable characteristics. It has poor solvent resistance and is slow to crystallize, which causes poor injection molding performance. For commercial blends, the transesterification reaction must be controlled [7, 8].

There are many other commercial examples of polymer blends. Polycarbonate can be blended with an acrylonitrile-butadiene-styrene terpolymer to give a PC-ABS blend. Polypropylene impact can be improved by the addition of ethylene-propylene copolymers, which are sometimes called ethylene-propylene-rubber (EPR). Ethylene, propylene, and a diene monomer (EPDM), such as ethylidene norbornene, is also used to impart impact and flexibility to polypropylene.

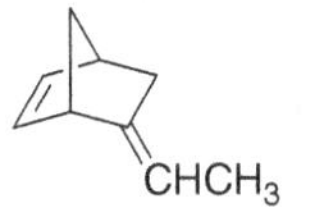

CHCH3
Ethylidene norbornene

There are numerous polyolefin blends. For example, blending a high molecular weight HDPE with LLDPE gives a blend useful for making thick films with high modulus (stiffness) and good tear strength [9]. Polypropylene blended with a maleic anhydride graft ethylene copolymer has improved adhesion to metals [10]. Ethylene ethyl acrylate (EEA) and ethylene vinyl acetate (EVA) have been added to polyolefins. They can improve properties such as surface adhesion for printability. When added to LLDPE, the toughness and elasticity of the LLDPE is improved [11].

9.2 ANTIOXIDANTS

Whether the plastic article is fabricated from a single polymer or a blend of polymers, it is subject to oxidative degradation. The susceptibility to oxidation will vary based upon the polymer and the other components in the formulation. For example, polypropylene which can form tertiary radicals is more susceptible than HDPE. Some additives can act as catalysts and accelerate degradation. The oxidation is a radical reaction. The initial steps are illustrated with polyethylene, but other polymers have a similar mechanism.

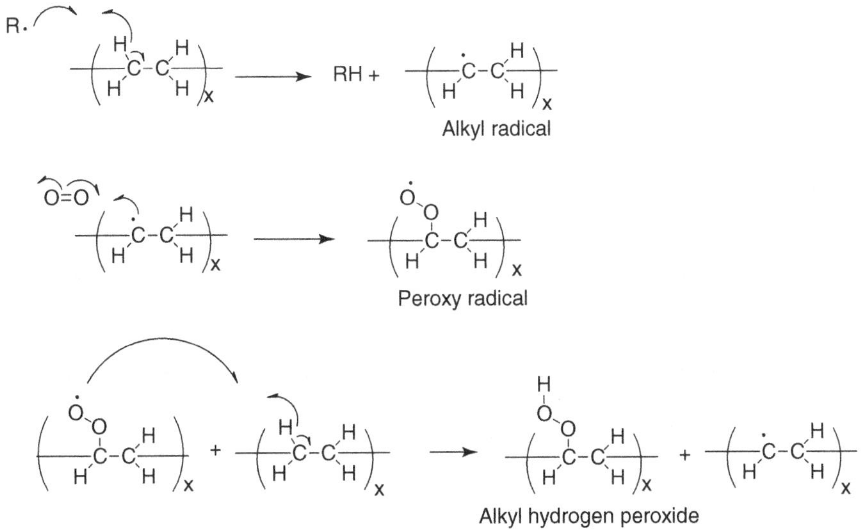

RH +
Alkyl radical

Peroxy radical

Alkyl hydrogen peroxide

This is only the beginning and both alkenes and carbonyl containing compounds are eventually formed. Severe oxidation of polyethylene, as might be observed in accelerated testing, can be monitored by FTIR analysis and the appearance of absorption in the carbonyl region [12].

Additives to stabilize typically work on one of three principles. Most commonly, they are radical scavengers. They react with a radical to form a non-reactive species. Another class of additives sacrificially reacts with oxygen. A third type will work by sequestering metals that can catalyze the degradation. Recognize that in the process of forming a plastic article, the polymer has been subjected to high heat and shear. Often the polymer has been manufactured in a thermal process. Typically it is then extruded with other additives. In the extrusion process it is subjected to a high enough heat to allow the polymer to flow. Also the turning of the screw causes localized frictional heating. Finally, the polymer is then melt-fabricated to form the final article. In each instance, there is an opportunity to break bonds and form radicals.

A common type of radical scavenger is a hindered phenol, such as butylated hydroxy toluene (BHT). A radical can react with BHT by removing a hydrogen atom and forming a new phenolic radical. However this radical is sterically hindered from further reactions by the two t-butyl groups and therefore the propagation steps terminate with the formation of this stable radical. Antioxidants that act as radical scavengers are sometimes referred to as primary antioxidants.

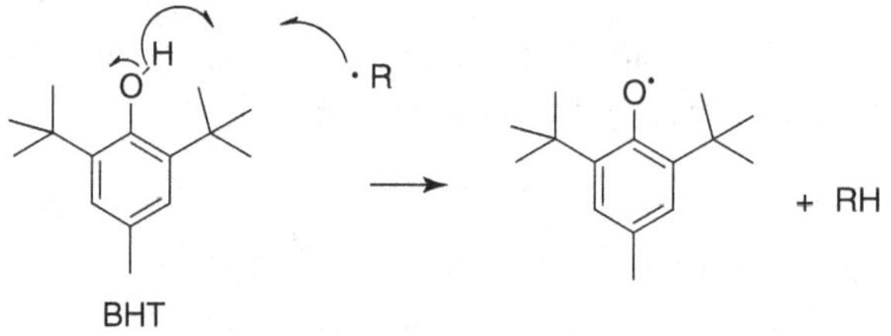

BHT and similar compounds are effective antioxidants and they have been used in polymers and also in food. However, in polymers, the BHT has enough volatility that it can come out of the polymer. When a material migrates out of the polymer, it is said to "plate out." This can happen during the molding step and BHT can plate out onto the mold as a powder. It can also happen gradually on storage and over time the article develops a white powder of BHT on the surface. The tendency to plate out also depends upon the solubility of the additive in the polymer. Plate-out is undesirable and therefore less volatile analogues of BHT are typically used. One common example is octadecyl 3,5-di-t-butyl-4-hydroxyhydrocinnamate, originally marketed by Ciba-Geigy Corp. as Irganox® 1076. As this and other additives came off patent and other companies began producing the additives, many of them continued with the same numbers, so this can be referred to as antioxidant 1076 or AO 1076. Other common antioxidants are AO 1010 and AO 1098, but there are many others based upon hindered phenols. They are selected based upon the polymer being stabilized and the desired properties such as color retention and extraction resistance. For example, AO 1076 and AO 1010 are used in several polymers including polyolefins, but for nylons AO 1098 is recommended.

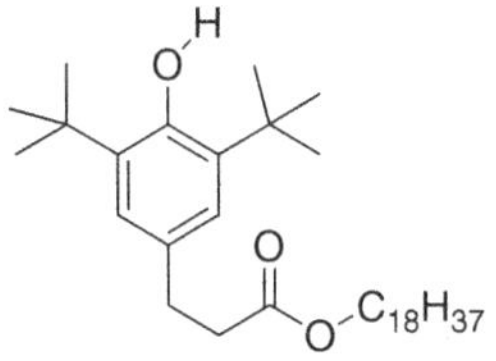

AO 1070

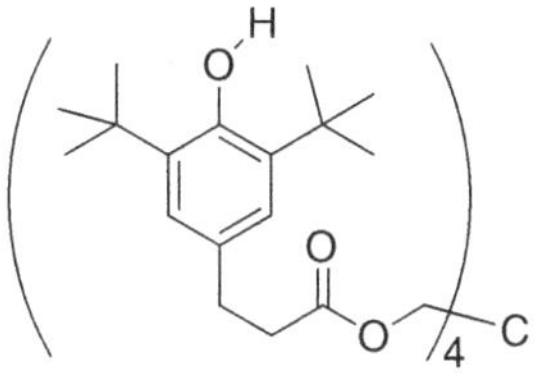

AO 1010

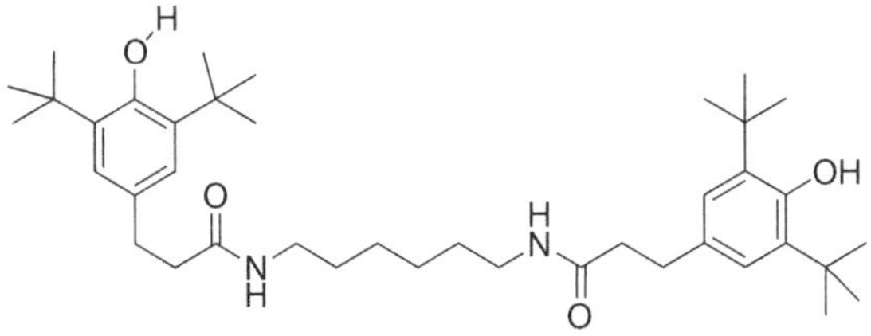

AO 1098

Amine antioxidants are also used. They too are considered to act as radical scavengers. Commonly these are substituted p-phenylenediamines or alkylated diphenyl amines. These are used more in rubber or lubricant formulations and less so in thermoplastics. Often they impart color so they are more likely to be used in formulations that are pigmented with carbon black.

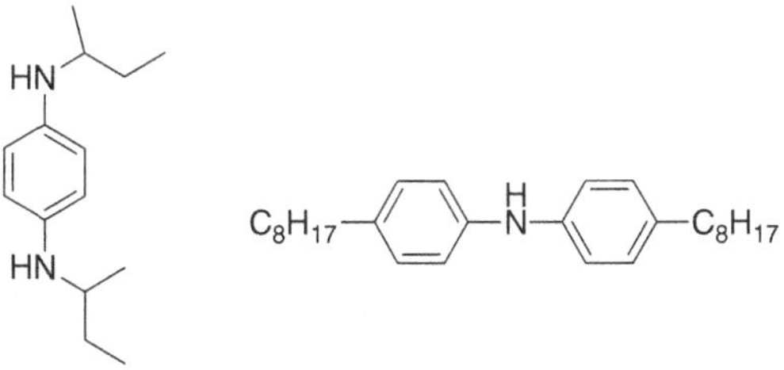

Amine antioxidants

Another type of antioxidant sacrificially reacts with oxygen, probably in the form of a peroxide. This type of antioxidant is often called a secondary antioxidant. Commonly they are used in conjunction with a primary antioxidant such as a hindered phenol. They are typically phosphites, which upon reaction with oxygen become phosphates or sulfides that can react with oxygen to form sulfoxides or sulfones, depending on the degree of oxidation. The same concerns about polymer compatibility and volatility apply here also.

$(RO)_3P \longrightarrow (RO)_3P{=}O$

Phosphite Phosphate

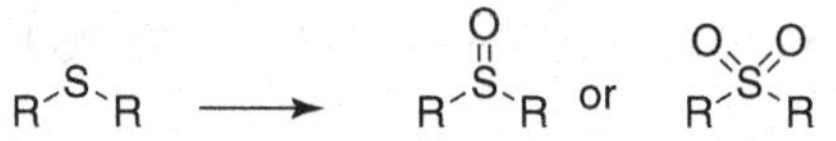

R–S–R ⟶ R–S(=O)–R or R–S(=O)(=O)–R

Sulfide Sulfoxide Sulfone

Tris(nonylphenyl)phosphate (TNPP) is a common commercial phosphite. Note that this is not n-nonyl, but rather a mixture of branched C9 isomers. Nonylphenol is commercially synthesized through Friedel–Crafts alkylation of phenol with nonene. Technical nonene ("propylene trimer") is a mixture of predominantly C9-olefins with varying degrees of branching. The resulting nonylphenol is mainly a mixture of 4-substituted monoalkylphenols with various isomeric, branched nonyl groups [13]. Tris(2,4-di-t-butylphenyl)phosphite is another commonly used phosphite.

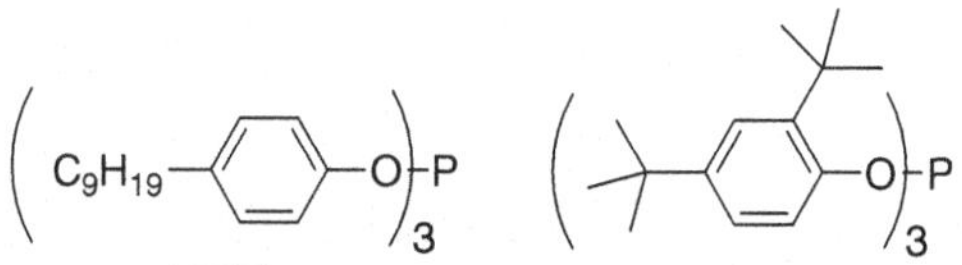

$(C_9H_{19}–C_6H_4–O)_3P$

TNPP

Common sulfides are distearylthiodipropionate (DSTDP) and dilaurylthiodipropionate (DLTDP). Although it is more proper to call these sulfides, they are often called thioesters.

$S(CH_2CH_2CO_2C_{18}H_{37})_2$ DSTDP

$S(CH_2CH_2CO_2C_{12}H_{25})_2$ DLTDP

It is well known that certain metals can catalyze oxidation reactions. Trace metals can be present in a polymer article from many sources including other additives, residual catalyst, or contamination from the processing equipment, which is metal. A class of antioxidants is metal deactivators. Their purpose is to chelate or passivate the metal rendering it ineffective as an oxidation catalyst. Irganox® MD 1024 is an example of a metal deactivator.

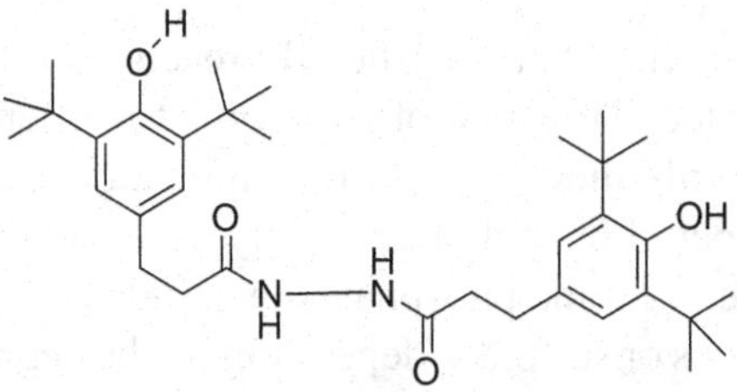

MD 1024

9.3 UV STABILIZERS

Many organic compounds, and therefore many polymers, are susceptible to reactions that are promoted by exposure to ultraviolet light. In a polymer, this can lead to color change and embrittlement. If a polymer absorbs UV radiation, the absorbed energy can cause homolytic bond cleavage resulting in radicals. As in the discussion above about oxidation, the alkyl radicals can react with oxygen to form peroxy radicals and the peroxy radicals can abstract a hydrogen atom from a polymer chain to form a hydroperoxide and another alkyl radical. Especially for articles that have outdoor exposure, some form of UV stabilizer is often needed. One type of UV stabilizer operates by absorbing the UV energy and then dissipating it by tautomerization. Examples of this type of stabilizer include the hydroxybenzophenones and the hydroxybenzotriazoles.

Tautomerization

OC_8H_{17}

Typical hydroxybenzophenone stabilizer

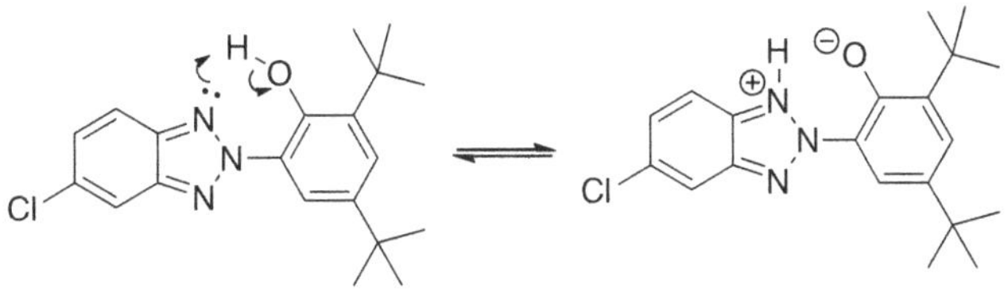

Typical benzotriazole stabilizer

This harmless dissipation of the UV energy stabilizes the polymer. Another type of UV stabilizer is a class known as hindered amine light stabilizers (HALS). Many of them are based upon tetramethylpiperidines.

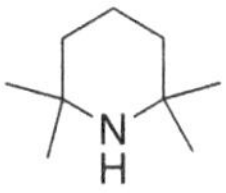

Tetramethylpiperidine

Some specific examples are shown with their original Ciba-Gcigy Corp. names, but they are sold under other names by other companies.

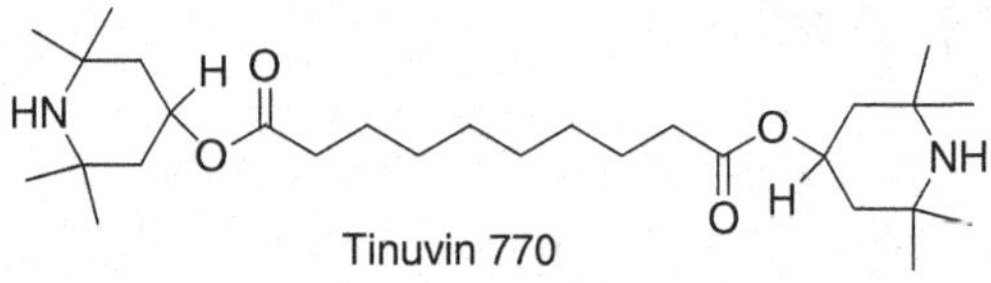

Tinuvin 770

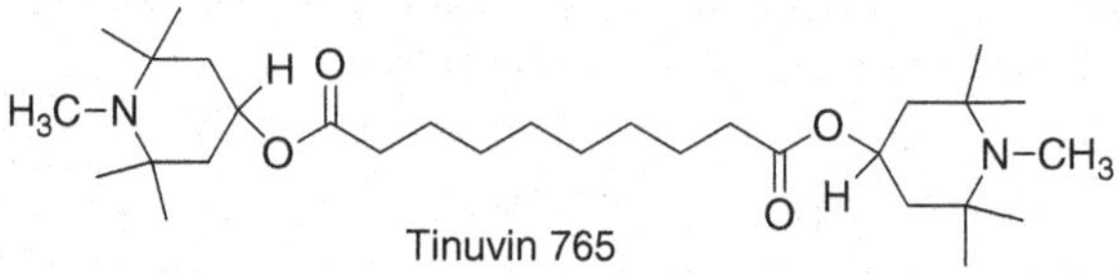

Tinuvin 765

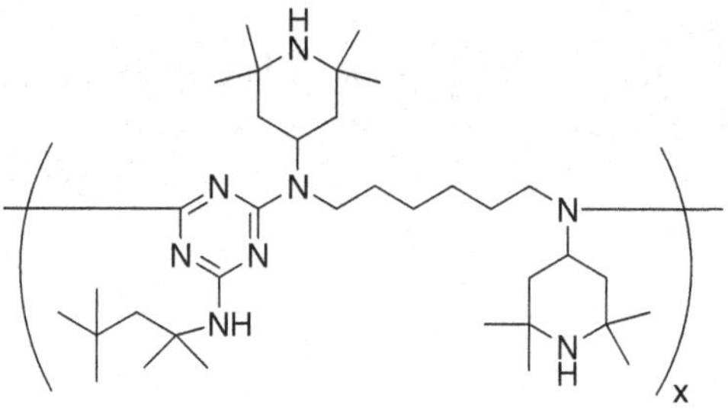

Chimassorb 119

HALS can be very effective. One explanation of their effectiveness is that the hindered amine oxidizes to form a nitroxyl radical, which can react with a polymer radical to form an alkoxy amine. The alkoxy amine can terminate peroxy radicals and in the process regenerate the nitroxyl radical.

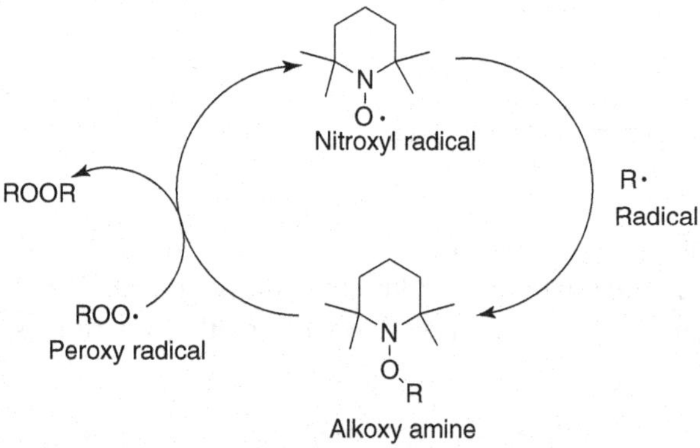

Nitroxyl radical

ROOR

R·
Radical

ROO·
Peroxy radical

Alkoxy amine

9.4 ANTISTATIC AGENTS

Antistatic agents are added to aid in the prevention of static build up and sparking in the processing and usage of final parts. This can be critical to part performance and sometimes safety. Antistatic agents dissipate surface charges which form during manufacture and use. Another advantage to the use of antistatic agents is that dust particles are attracted to a lesser extent, and the dust buildup on the plastic article is reduced. For some applications, especially transparent articles which rely on optical properties, dust can be a problem. Migrating antistats diffuse to the polymer surface over time, creating a thin layer that attracts water molecules. The water molecules provide a conductive pathway that prevents build-up of static electricity and reduces the polymer's surface resistivity [14]. There are many additives that function as antistatic agents. Examples include glycerol monostearate, ethoxylated fatty acid amines such as ethoxylated stearylamine, and diethanolamides such as diethanol stearamide.

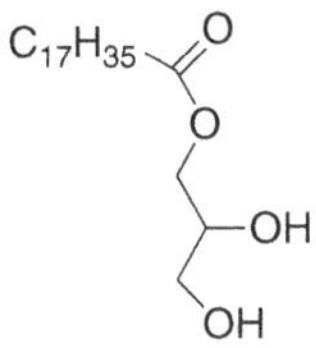

$C_{17}H_{35}$ $C_{18}H_{37}$ $C_{17}H_{35}$

Glycerol monostearate 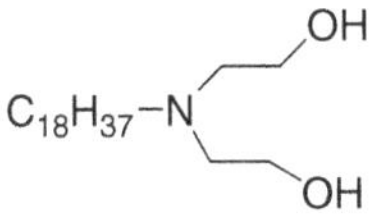 Ethoxylated stearylamine 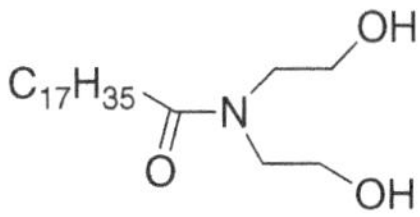 Diethanol stearamide

For some applications, such as trays of conveyer belts for electronic components, copiers and printers, or even artificial turf, more permanent antistats are desired. One approach is the use of polymers such as polyether amide copolymers or ethylene ionomers. Highly conductive graphite and metal fibers such as nickel fiber are also used to lower the resistivity of polymer articles [15]. For high performance materials that can tolerate the added cost impact, carbon nanotubes can also be used.

9.5 PEROXIDES

Peroxides have many uses in polymer manufacture. They are used as initiators in radical polymerization processes such as the formation of PVC, polystyrene, and low density polyethylene. They can also be added in a polymer formulation for various reasons. Sometimes this is to crosslink an unsaturated polymer such as unsaturated polyesters used in fiberglass formulations. Other times it is to graft one polymer to another chemical or polymer by a radical reaction. The choice of the peroxide is dictated by considerations such as storage stability, solubility, and decomposition to form radicals at a temperature near the desired reaction. Common peroxides include benzoyl peroxide, di-t-butylperoxide, and dicumylperoxide (DICUP).

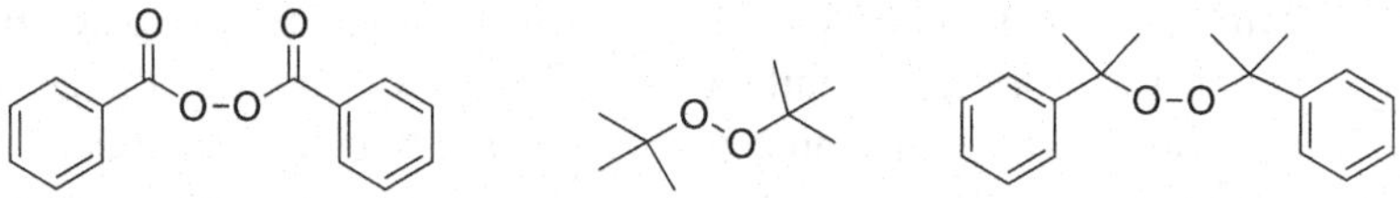

Benzoyl peroxide Di-t-butylperoxide Dicumylperoxide

Peroxides can be used to promote crosslinking of a rubber as is done in dynamic vulcanization. Dynamic vulcanization is usually done in an extruder and is the process of selectively crosslinking rubber during its melt mixing with a molten thermoplastic such as polypropylene. This produces thermoplastic vulcanates, often referred to as TPVs [16]. Peroxides are also used to provide grafting reactions. For example, polypropylene, maleic anhydride, and a peroxide initiator in a twin-screw extruder give a graft polypropylene [17]. These polymers can have improved properties such as improved adhesion.

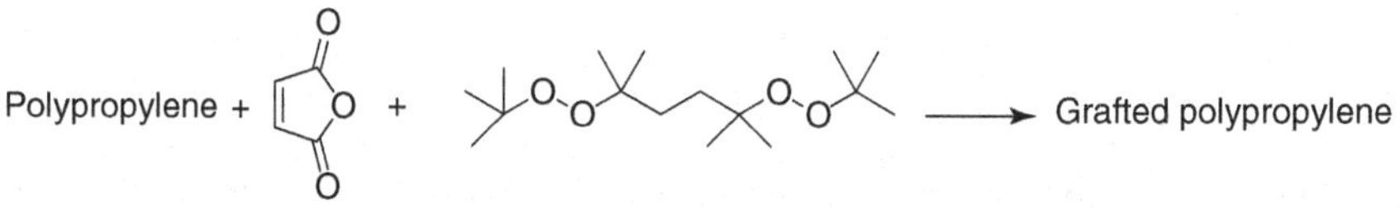

Polypropylene + + → Grafted polypropylene

Polyethylene can be crosslinked with peroxides to make it more suitable for coatings for electrical power cables [18]. Peroxides are also used to crosslink (also called cure) thermoset polyesters. Thermoset polyesters have a reactive alkene and are used to make fiberglass compositions such as used for boat hulls. Variations are possible, but commonly they are copolymers of fumaric acid or maleic anhydride with phthalic acid or isophthalic acid and a diol. The oligomer is typically dissolved in styrene. When a peroxide is mixed with the solution, crosslinking occurs and the polyester cures. The styrene solvent takes part in the reaction and the solution hardens.

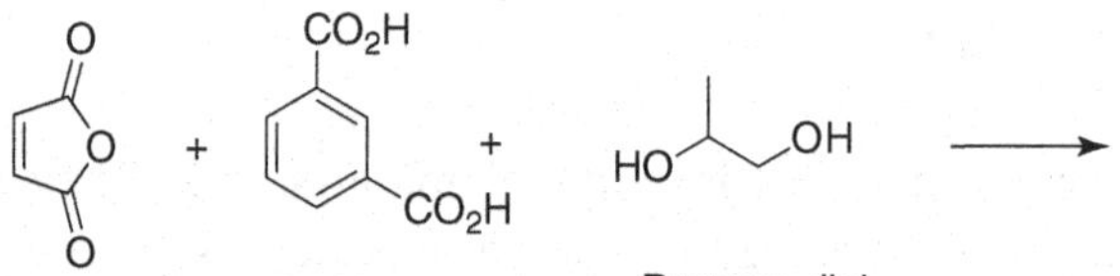

Maleic anhydride + Isophthalic acid + Propane diol →

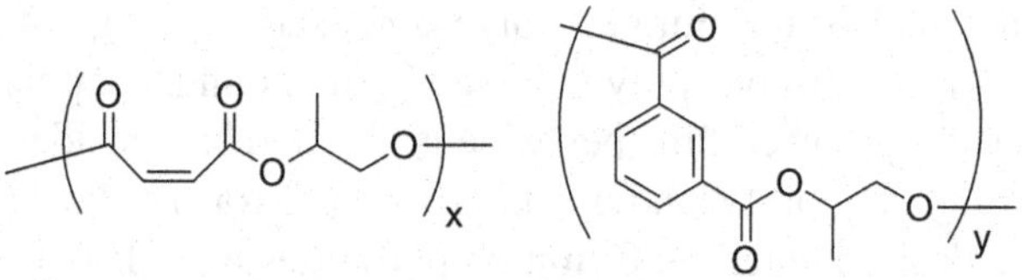

Oligomeric unsaturated polyester copolymer

9.6 LUBRICANTS

Lubricants are sometimes added to polymer formulations. Lubricants minimize frictional forces between moving surfaces. They can be used to improve processability or to improve dispersion of other additives such as pigments or fillers. They can be external lubricants to provide lubrication between the polymer surface and metallic processing equipment such as molds used in injection molding. The external lubricant can be applied to the mold surface as a spray or it can be added to the polymer formulation where it migrates to the polymer surface. Commonly these are paraffin waxes (oligomeric polyethylene), fatty acids such as stearic acid, fatty acid salts such as sodium stearate, or fatty acid amides, such as stearamide. One issue with external lubricants is that they migrate to the surface; that is, they plate out on the surface. Internal lubricants typically have greater solubility in the polymer. They promote movement of polymer chains in the bulk polymer. Like plasticizers they can lower the heat properties of the formulation, but unlike plasticizers, they are typically used in low levels to minimize the effect on upper use temperature of the molded article and on modulus. Esters of fatty acids such as glycidyl ester or butyl stearate are used. Another example is fatty acid amides such as ethylene bis-stearamide.

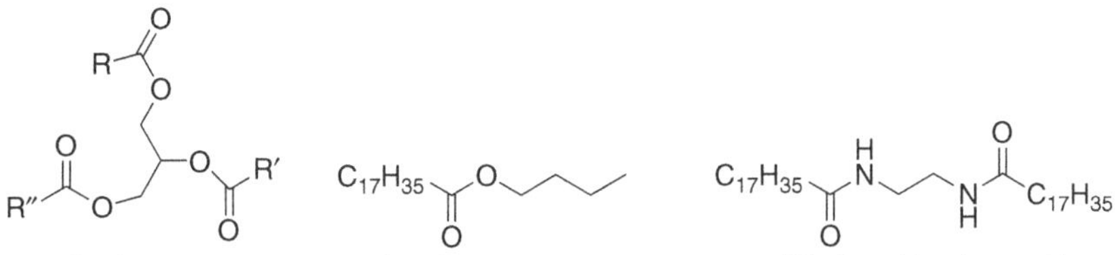

Glycidyl ester | Butyl stearate | Ethylene bis-stearamide

9.7 FLAME RETARDANTS

Flame retardants are additives used to retard the ignition and burning of plastics. By the use of flame retardants, plastic formulations can meet the various fire safety standards needed. One agency that provides fire resistance certifications is Underwriters Laboratory Inc. These certifications are often in the form of a UL rating. There are various UL tests and ratings. The goal is to make the test relevant to the application. One common way of rating plastics is the UL 94 vertical burn test. A flame is applied for 10 seconds to a plastic specimen and the removed until the flaming stops. The flame is then reapplied for another 10 seconds and removed. Depending on how long the specimen burns and whether there are flaming plastic drips, a rating is assigned.

To understand how flame retardants work, it is useful to think about the fire triangle. The fire triangle is a visual aid to show that heat, fuel, and oxygen are required for combustion. If any one of these is removed, combustion is not maintained. Consider an unlit candle in a jar. Fuel (the candle wax) and oxygen (in the air) are present but there is no combustion unless heat is supplied, initially in the form of a match and subsequently from the burning vapor of the degraded candle wax. Combustion is maintained until one of the three elements of the fire triangle is removed. When all the candle wax has been consumed, there is no more fuel and the flame goes out. If, while the candle is burning, a cover is put over the jar, then no more oxygen is supplied and the flame goes out.

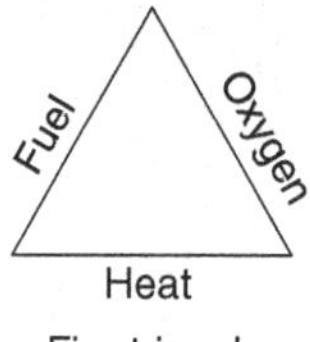

Fire triangle

Many plastic articles require flame retardants so that adequate safety is provided. Some areas where they are used include the manufacturing of mattresses and upholstery, in electrical applications such as wires and cables, in building construction, and in transportation such as aircraft and automotive interiors. There are three different mechanisms that can explain how most flame retardants work. One is by absorbing heat by the release of water. A second type of flame retardants is thought to work by providing an insulating char on the burning plastic. This char prevents more fuel (plastic) from reaching the flame and starves the fire. A third method is by interfering with the chemical reactions that maintain the fire and promote the spread of the flame.

The largest volume flame retardant is alumina trihydrate, $Al_2O_3 \cdot 3H_2O$. In the flame, water is released and this cools the plastic, preventing further combustion. Alumina trihydrate is used in polymers such as polyolefins, PVC, polyacrylates, and thermoset polyesters. It cannot be used in polymers such as polycarbonate, nylon 6,6, or polyethylene terephthalate because these are processed at temperatures that will cause the evolution of water during processing.

Phosphorus-based flame retardants such as ammonium polyphosphate or phosphate esters are thought to form phosphoric acid in the flame and the acid is thought to promote char forming crosslinking reactions. They are often used in polyurethanes and in polyphenylene oxide. A type of flame retardant useful for polycarbonate are salts of sulfonic acids [19, 20] such as the potassium salt of aromatic sulfonic acids, exemplified by potassium toluene sulfonate. These are thought to promote a char layer in the flame.

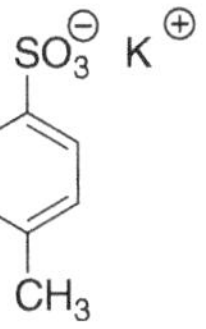

$SO_3^{\ominus}$ $K^{\oplus}$

CH_3

Potassium p-toluene sulfonate

A third type of flame retardant interferes with the chemical reactions that maintain the combustion. These are based upon halogenated compounds, often brominated aromatic compounds. These flame retardants are somewhat thermally labile and release bromine radicals that quench the radical chain reactions during the combustion and flame spreading processes. Antimony oxide, Sb_2O_3, is often added because it has a synergistic effect with the brominated flame retardant. Antimony oxide increases the rate of release of halogens via the formation of antimony halides and oxyhalides during combustion [21]. The bromine radical abstracts a hydrogen atom to form HBr, which can then quench hydrogen radicals, forming hydrogen and a bromine radical. HBr can also quench a hydroxyl radical, forming water and a bromine radical. The hydrogen radical is responsible for the chain-branching free radical reactions in the flame and the hydroxyl radical is responsible for the oxidation of carbon monoxide to carbon dioxide which is a highly exothermic reaction and responsible for the larger part of the heat generation in the flame [22].

Because of the synergistic effect of antimony oxide, it is almost always added with the halogenated flame retardant into the formulation. All halogenated compounds can act as flame retardants, so PVC with added antimony oxide can be flame retardant. However, aliphatic halogenated compounds are less common than aromatics because of their ability to undergo dehydrohalogenation during processing, especially when used for polymers that are processed at high temperatures. Brominated aromatic compounds are stable during processing but readily decompose in the flame. The more thermally stable chlorinated aromatic compounds are slower to decompose in the flame and are less effective in most applications. Some brominated flame retardants are shown.

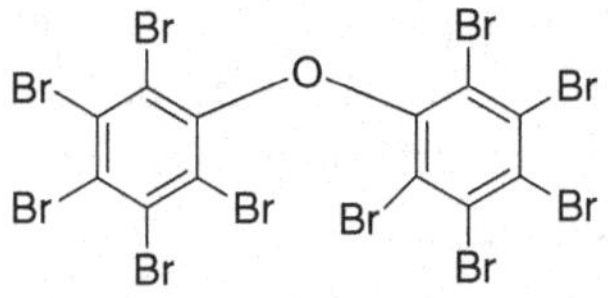

Decabromodiphenyl ether

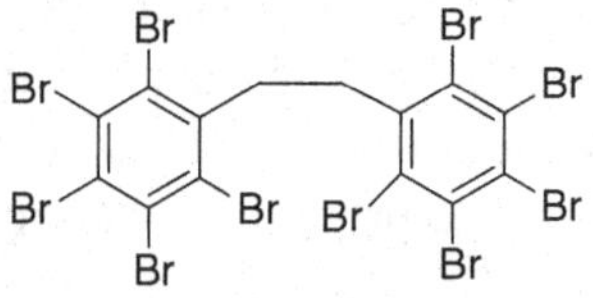

Decabromodiphenylethane

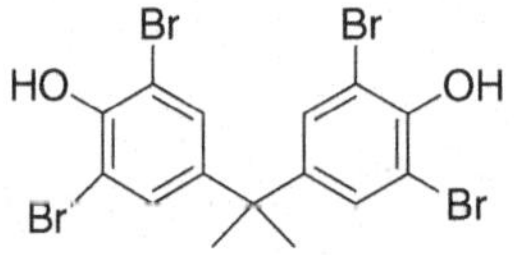

Tetrabromobisphenol A

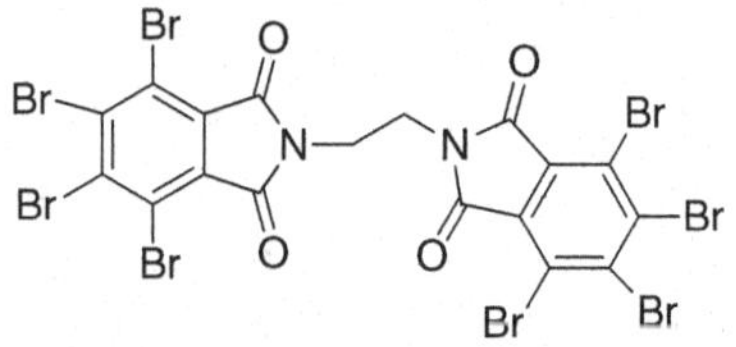

Ethylene bis-tetrabromophthalimide

Decabromodiphenyl ether is being phased out because of concerns about the effect on human health and the environment. There is also a concern that under some waste incineration conditions that might be used for discarded plastics, it can be converted to polybrominated dioxins which bear a structural similarity to 2,3,7,8-tetrachlorodibenzo-p-dioxin (TCDD).

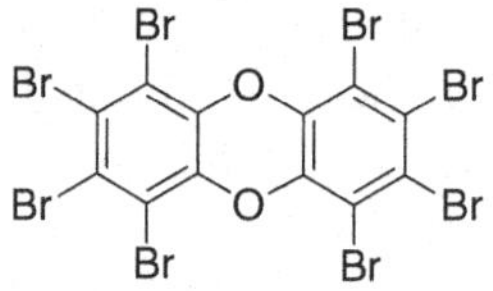

A polybrominated dioxin

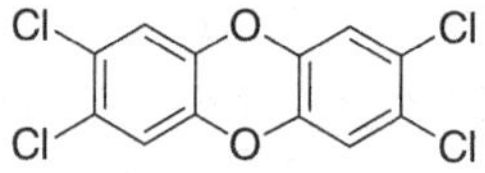

TCDD

One chlorinated flame retardant that is useful for high processing polymers such as nylon is Dechlorane Plus®. This has good thermal stability and is not prone to dehydrochlorination because the reaction would involve formation of double bonds at bridgehead carbons.

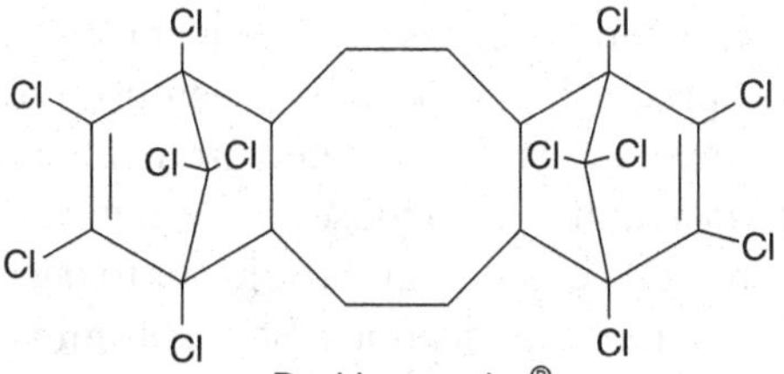

Dechlorane plus®

Antimony oxide is prepared from naturally occurring stibnite ore. Stibnite is Sb_2S_3. The antimony ore (stibnite) is roasted to vaporize the stibnite which is then oxidized by the introduction of air followed by controlled cooling of the gas.

$$2\ Sb_2S_3 + 9\ O_2 \longrightarrow 2\ Sb_2O_3 + 6\ O_2$$

Antimony oxide can also be prepared from the hydrolysis of antimony trichloride. Antimony trichloride is prepared by the reaction of calcium chloride with stibnite and oxygen.

9.8 HEAT STABILIZERS

PVC is an important and widely used polymer but it has poor thermal stability. With heat or light, it can undergo dehydrochlorination reactions, forming alkenes.

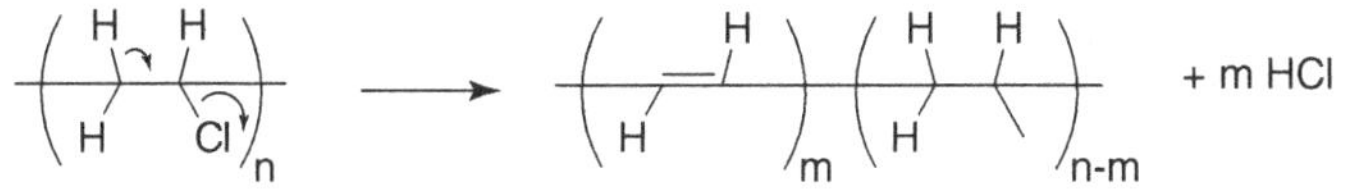

H H H Cl n → H H m H H H n-m + m HCl

This reaction is sometimes called unzippering and is thought to originate at certain structural defects within the polymer [23].

Nine structural defects have been shown to occur in all samples of commercial PVC and they include some structures which contain tertiary chlorines as seen in two of the identified defect structures.

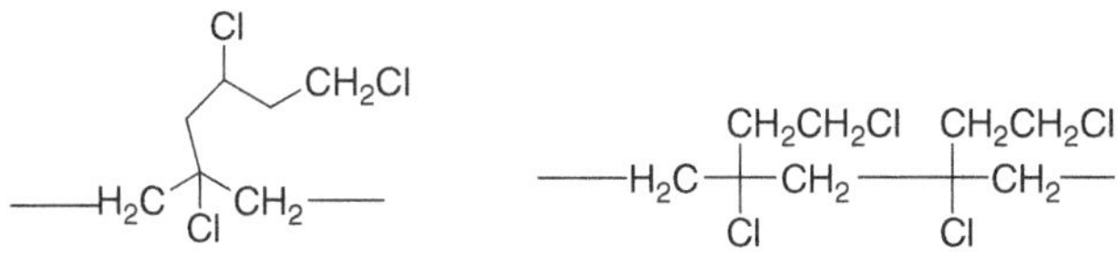

Cl CH_2Cl —H_2C Cl CH_2— CH_2CH_2Cl CH_2CH_2Cl —H_2C—CH_2——CH_2— Cl Cl

Defect structures found in PVC

When the PVC unzippers, it discolors and becomes brittle. To prevent this dehydrochlorination reaction, heat stabilizers are added. Lead stabilizers are effective and a major type of heat stabilizer. Data from 2003 [24] shows lead stabilizers to represent more than 50% of the world PVC stabilizers, but less than 20% in the U.S. where tin stabilizers predominate. The use of lead stabilizers is on the decline and being phased out due to concerns about the health effects of lead. Nonetheless, the news occasionally reports that lead has been found in a toy or other PVC article imported from a country such as China. The source of lead is likely due to the stabilizer. Another type of heat stabilizers is the mixed metal packages which are based upon mixtures of different metals such as barium, cadmium, calcium, and zinc, often in the form of fatty acid salts such as zinc stearate.

Tin stabilizers are commonly used. There are many types in use but generally they are monoalkyl or dialkyl tin esters or mercaptides. Common alkyl groups are methyl, butyl, and octyl. Alkyl tin thioglycolates are effective heat stabilizers and are commonly used.

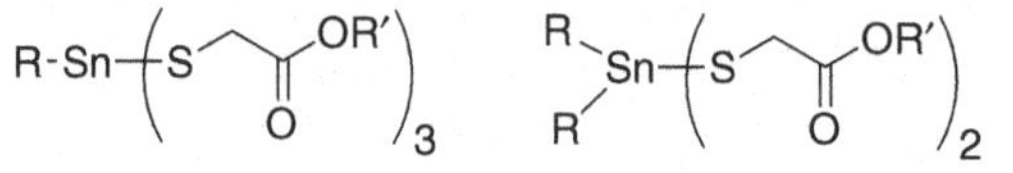

Alkyltin thioglycolate Dialkyltin thioglycolate

9.9 PLASTICIZERS

Plasticizers are compounds added to polymers to facilitate processing and to increase the flexibility and toughness of the final product. The plasticizer reduces the intermolecular attractions between polymer chains. The overwhelming majority of applications impart flexibility to PVC. The most common plasticizers are aliphatic esters of various carboxylic acids or aromatic esters of phosphoric acid.

Examples of common plasticizers include phthalate esters such as dioctyl phthalate (DOP), more properly called di-2-ethylhexyl phthalate (DEHP), and used as an inexpensive general purpose plasticizer. This is made by reacting phthalic anhydride with 2-ethylhexanol. Another example is diisodecyl phthalate (DIDP). DIDP has lower volatility and improved resistance to soapy water extraction. It has many applications in PVC used for wire and cable coating. Phthalate plasticizers are also made from linear alcohols such as 1-hydroxyheptane, 1-hydroxynonane, and 1-hydroxyundecane. They are often used when superior low temperature properties, lower volatility, or outdoor weathering is required.

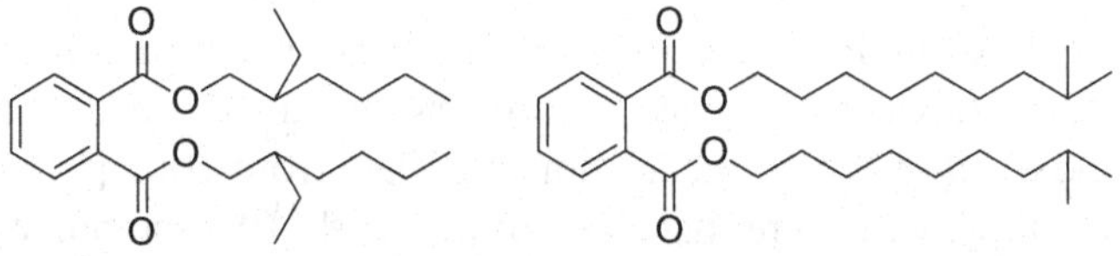

DOP DIDP

Over long periods of time or at higher temperatures the plasticizer can volatilize from the plastic. Upon aging, PVC can lose its flexibility and embrittle because of the fugitive nature of plasticizers. Compared with phthalates, plasticizers based upon trimellitic anhydride offer greater permanence at elevated temperatures. Trioctyl trimellitate (TOTM) is much less volatile than DOP, perhaps by about two orders of magnitude under some circumstances. This can be important for applications where building codes require higher temperature rating of electrical wiring.

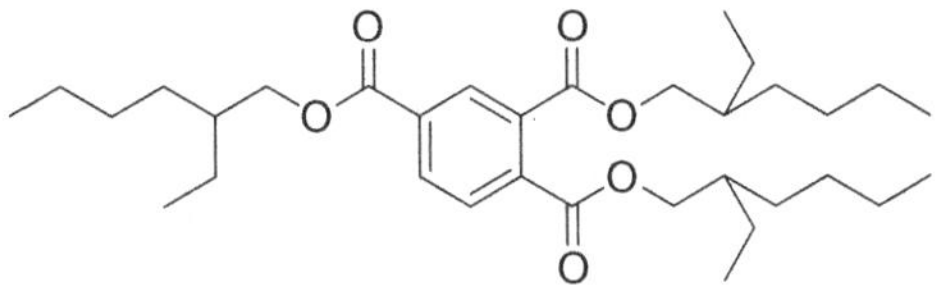

TOTM

Plasticizers based upon aliphatic dicarboxylic acids are also used, mainly where low temperature flexibility of PVC is required such as in some film and plastic wrap applications. One common plasticizer in this class is dioctyl adipate (DOA).

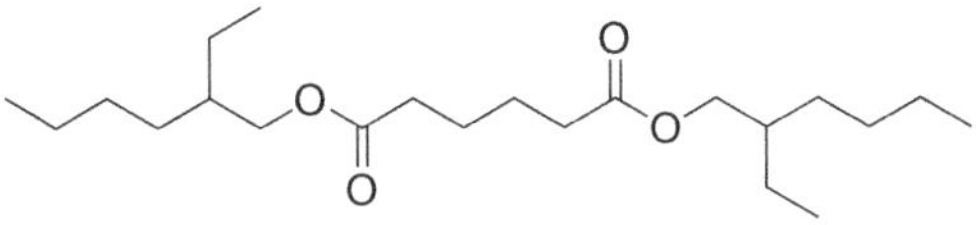

DOA

Aliphatic dicarboxylic acids are also esterified with glycols such as 1,4-butanediol to make polymeric plasticizers. They have low volatility and resistance to migration and extraction.

Aromatic esters of phosphoric acid are also sometimes used. They can have a plasticization effect but can also act as flame retardants. They are used in some specialty PVC applications, cellulose acetate film applications, and polyphenylene oxide blends. Examples include triphenyl phosphate and resorcinol bis(diphenyl phosphate).

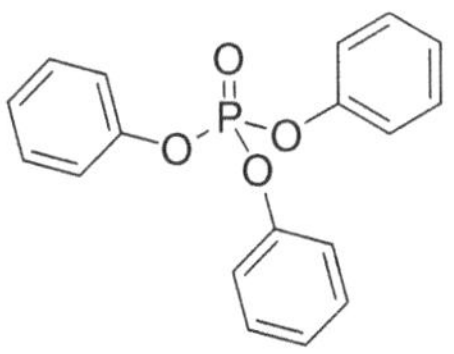

Triphenyl phosphate

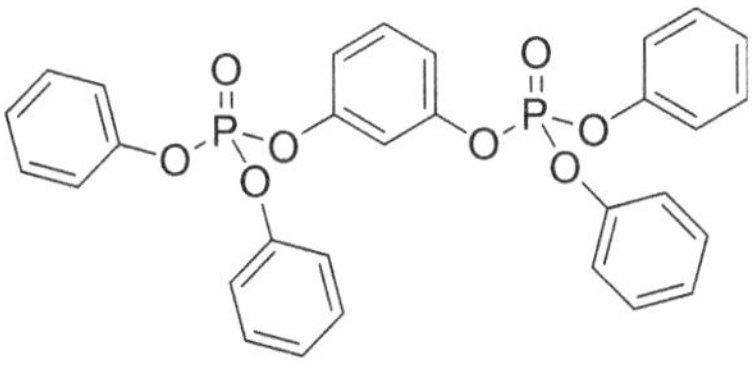

Resorcinol bis(diphenylphosphate)

9.10 OTHERS

There is a myriad of other additives commonly used in polymer formulations. Inorganic fillers such as mica, clay, or glass fibers add strength and stiffness to the polymer article. They can also decrease mold shrinkage. When a plastic article is formed by injection molding, upon cooling it shrinks and the dimensions are slightly smaller than those of the mold. This is because the density of a polymer increases as the temperature decreases. Shrinkage is greater in crystalline polymers because with cooling as the crystallites form, the density increases. Shrinkage can cause sink marks, voids, and residual stresses and needs to be controlled. In extreme situations, the plastic part can warp or crack due to the stresses. Some shrinkage can be controlled by injection molding conditions and mold design. If the shrinkage is consistent, the size difference can be compensated for by making the mold cavity appropriately larger when the mold is first made. It is more of an issue when different formulations with different amounts of shrinkage are processed in the same mold. For many materials, there can be post-mold shrinkage as the article slowly anneals.

Impact modifiers are added to many formulations. As the name implies, they impart toughness to the polymer article or film. Many of them are butadiene copolymers that disperse in the polymer matrix. One type – the so-called core shell modifiers – has a rubbery core surrounded by a harder acrylate layer. They have been compared to an egg; soft on the inside and hard on the outside. The outer shell also has some adhesion to the matrix so that the modifier can be dispersed. Other impact modifiers include methacrylate-butadiene-styrene copolymers or EPDM, ethylene-propylene-diene monomer copolymers. Acrylonitrile-butadiene-styrene (ABS) and ethylene-vinyl acetate (EVA) are also used.

Pigments are also common. Generally when a formulation is pigmented it contains several pigments to give a color match. Carbon black is added to make black formulations; white formulations are made with titanium dioxide. For other colors, titanium dioxide is often added to provide a white base so that minor differences in the yellowness of the polymer do not prevent a good color match.

Some formulations contain nucleating agents to speed the crystallization of the polymer. Agents such as sodium stearate or calcium stearate are used. One type of nucleating agent is 1,3-O-2,4-bis(3,4-dimethylbenzylidene) sorbitol [25] which provides good clarity for polypropylene.

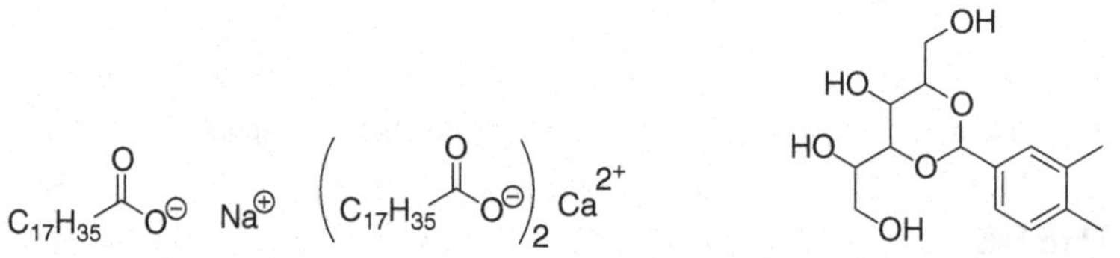

Sodium stearate   Calcium stearate   1,3-O-2,4-bis(3,4-dimethylbenzylidene) sorbitol

Plastic articles or films that come in contact with food might be formulated with antimicrobial additives. Other applications that might use antimicrobial additives are those that are used in a wet environment such as in the bathroom or pool liners. Inorganic antimicrobial agents can be used. Silver ions are effective antimicrobial agents and can be encapsulated to provide controlled release into the polymer matrix [26]. Organic antibacterial additives can also be used. Examples include triclosan and salts of the biguanide, chlorhexidine.

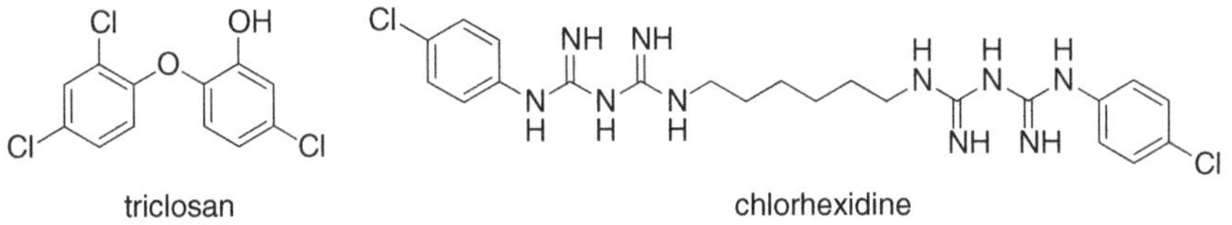

triclosan chlorhexidine

A typical plastic formulation might contain two or three polymers and five or six additives. During processing the formulation is often heated to high temperatures and thoroughly mixed during extrusion. Chemical reactions can take place and need to be considered by the formulator. When thinking of polymers, we often focus on the repeat units, but they have end groups. For example, a polyester can end in a carboxylic acid end group or in an alcohol end group. The polyester then might have three functional groups; ester, alcohol, and carboxylic acid as well as residual catalyst. Any of these can react with other functional groups present in other polymers or with additives. It is common for additives to react with other additives. For example, antimony oxide, present either as a residual catalyst or as a flame retardant additive, can react with a phosphite antioxidant additive and cause a black color. When this is a problem thioesters are used as secondary antioxidants instead of phosphites. Sometimes the additives can have a deleterious effect on aging properties. Other additives may bloom to the surface and cause poor adhesion in secondary operations such as painting. Proper selection of additives requires consideration about the end-use applications and an understanding of possible interactions with other ingredients in the formulation.

QUESTIONS

1. A miscible blend always
   - **a.** has a single Tg
   - **b.** has two Tgs
   - **c.** has a melting point and a Tg
   - **d.** has two melting points
2. This word is used to describe a situation where the combination is greater than the sum of the parts, for example when the combination of

a brominated flame retardant and antimony oxide performs better than either component.

a. sycophant
b. stupendous
c. syllogism
d. synergy
e. syllogistic

3. Label each chemical for its function. Choose from:
antioxidant
UV stabilizer
flame retardant
antistatic agent
lubricant
PVC heat stabilizer

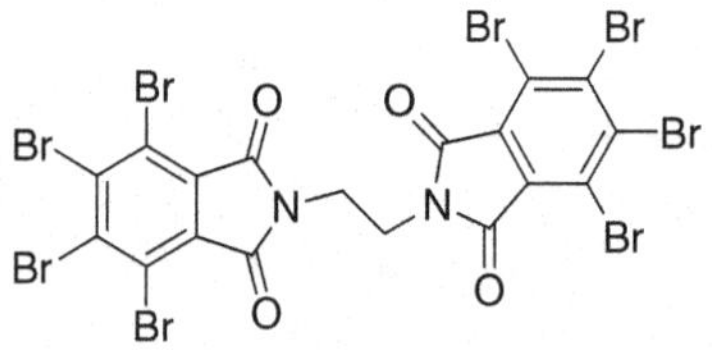

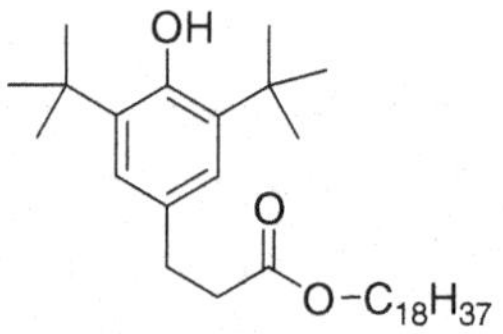

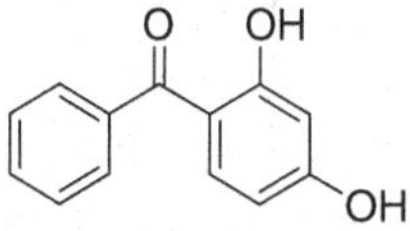

$C_4H_9\text{-}Sn\text{-}(SCH_2CO_2C_8H_{17})_3$ $C_{17}H_{35}$ OC_4H_9

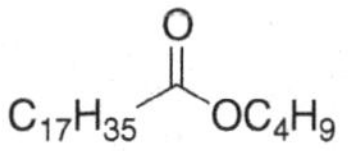

4. Choose the structure that is **not** commonly used as an antioxidant for polymers.

(a) $C_{18}H_{35}$ $C_{18}H_{17}$ (b) $C_{17}H_{35}$ NH_2

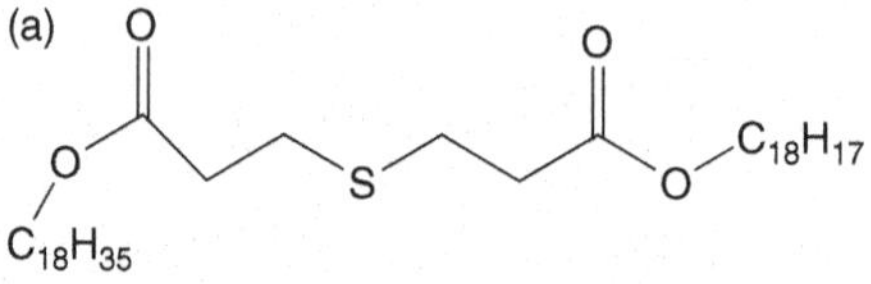

(c) C_9H_{19} (d) $CO_2C_{17}H_{35}$

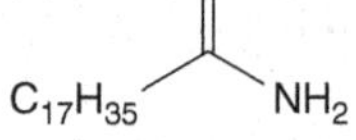

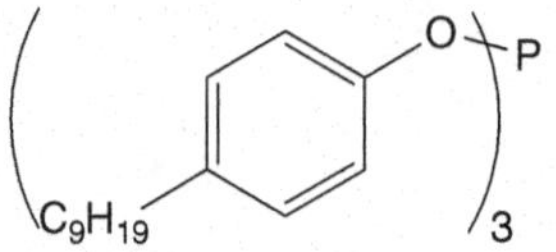

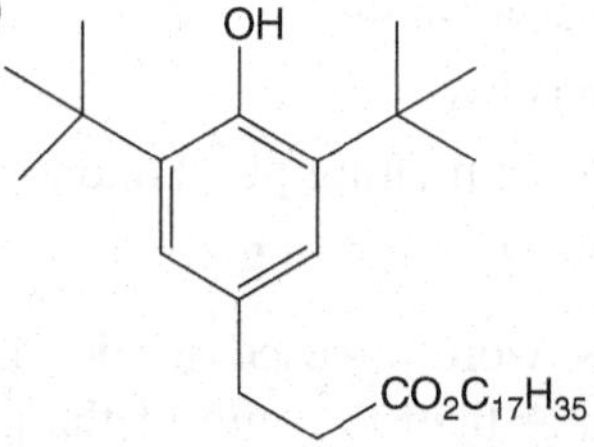

REFERENCES

1. Allan S Hay. U.S. Pat. No. 3,306,874. 1967.
2. Allan S Hay. U.S. Pat. No. 3,306,875. 1967.
3. Allan S Hay. U.S. Pat. No. 4,028,341. 1975.
4. Manas Chanda, Salil K Roy. *Plastics Technology Handbook*. CRC Press, Boca Raton, FL; 2006: 4–126.
5. S C Cohen, R Dieck. U.S. Pat. No. 4,257,937. 1981.
6. J Lohmeijer, J Heuschen. U.S. Pat. No. 4,654,400. 1987.
7. Donald BG Jaquiss, Russell J Mc Cready, John A Tyrell. U.S. Pat. No. 4,532,290. 1985.
8. John A Tyrell. U.S. Pat. No. 4,560,722. 1985.
9. Harilaos Mavridis. U.S. Pat. No. 7,608,327. 2009.
10. Christian Leboeuf. U.S. Pat. No. 7,550,533. 2009.
11. Mahmoud Rifi. U.S. Pat. No. 5,326,602. 1994.
12. Rui Yang, Ying Liu, Jian Yu, Kunhua Wang. Polymer Degradation and Stability 2006; 91:1651–1657.
13. Bjoern Thiele, Volkmar Heinke, Einhard Kleist, Klaus Guenther. Environ. Sci. Technol. 2004; 38:3405–3411.
14. Jennifer Markarian. Plastics Additives & Compounding September/October 2008; 10(5):22–25.
15. Jennifer Markarian. Plastics Additives & Compounding September/October 2008; 10(5):22–25.
16. Oansuk Chung. U.S. Pat. No. 7,625,980. 2009.
17. Vicki Flaris, David Mitchell. U.S. Pat. No.6,228,948. 2001.
18. Paul Caronia, Jeffrey Cogen. U.S. Pat. No. 6,656,986. 2003.
19. Victor Mark. U.S. Pat. No. 3,978,024. 1976.
20. Niles Rosenquist, John Tyrell. U.S. Pat. No. 4,535,108. 1985.
21. Guido Grausea, Jun Ishibashia, Tomohito Kamedaa, Thallada Bhaskarb, Toshiaki Yoshioka. Polymer Degradation and Stability 2010; 95(6):1129–1137.
22. Alexander B Morgan, Charles A Wilkie. *Flame Retardant Polymer Nanocomposites*. New York: Wiley-Interscience; 2007:9.
23. WH Starnes. J. Polym. Sci. Part A: Polym. Chem. 2005; 43(12):2451–2467.
24. Charles E Wilkes, James W Summers, Charles Anthony Daniels, Mark T Berard. PVC Handbook. Hanser Verlag, Cincinnati, OH; 2005:97.
25. Rob Hanssen. U.S. Pat. No. 7,879,933.2011.
26. Jeffrey Trogolo, Frank Rossitto, Edward K Welch II. U.S. Pat. No. 7,357,949. 2008.

CHAPTER 10

Pharmaceuticals

10.1 THE DRUG DEVELOPMENT PROCESS

It is a long process to bring a drug to market. Early experimentation leads to a compound that has potential to enter into clinical trials. This is often called the discovery phase and generally takes three to five years [1]. During discovery, leads are identified and optimized. Assays are developed and validation work is done. Often many structurally similar compounds are synthesized and studied. Next, preclinical studies are done to test the safety in animals, stability testing is done, and the drug metabolism is studied. The preclinical phase typically takes about one to two years. At this point, if everything looks promising, an Investigational New Drug application (IND) is filed with the FDA. The IND includes the animal toxicology and pharmacology studies, manufacturing information to ensure that the company can produce and supply consistent batches of the drug, and protocols for the proposed clinical studies.

The IND is the precursor to human studies. In Phase 1, healthy volunteers are tested. Phase 1 testing is for safety and dosage and typically takes about a year. Once the drug is deemed safe in healthy volunteers, Phase 2 begins. The drug is tested with people who have the condition being treated to determine effectiveness, side effects, and dosage. As the drug proceeds from Phase 1 to Phase 2 and then Phase 3, greater amounts of the drug need to be manufactured, increasing numbers of people are given the drug, and study costs increase. Phase 1 might test 20–100 healthy volunteers, Phase 2, 100–500 patients, and Phase 3 may have 1,000–5,000 patients in the study. Phase 2 might take a couple of years. If a drug continues to look promising, it enters Phase 3 of the drug development process. Phase 3 involves large-scale clinical trials to demonstrate safety and efficacy over longer periods of time. The new drug is compared against existing therapies and placebos. Phase 3 typically takes about two to five years.

Fundamentals of Industrial Chemistry: Pharmaceuticals, Polymers, and Business, First Edition. John A. Tyrell.

After a successful Phase 3 study, a company wishing to commercialize a new drug will file a New Drug Application (NDA). The amount of information required in a NDA is extensive. The shorter the time to review an application, without compromising safety, the sooner sick people can take advantage of the new drug. Incomplete NDAs slow the process. In an effort to expedite, the FDA has issued about 40 guideline documents to help sponsors with the process. The amount of time for review varies, but about one year is common. After commercialization, the work is not over. For example, since Lipitor's launch in 1997, Pfizer has invested over $800 million in dozens of studies involving over 50,000 patients [2].

The whole drug development process takes about 10–15 years. The cost to develop a new drug has been estimated at $1.3 billion [3] and only about 20% of new molecular entities (NMEs) cover their average capitalized R&D expenses [4]. Despite the long process and the high expense, the United States pharmaceutical industry spends about $65 billion each year on R&D and has about 3,000 compounds in development. As a drug candidate moves through the development process, costs escalate. From discovery through preclinical up to Phase 1 may cost about $30 million. Phase 1 studies are about $10–15 million per compound; Phase 2 about $60–100 million; and Phase 3 about $400–$800 million [5].

The drug development process is not only long and costly, but also high risk. A compound in discovery has about a one in 10,000 chance of commercialization. For those candidates that have passed preclinical and enter Phase 1 testing, there is less than a 12% success rate. Even those entering Phase 3 have a success rate of around 50% [6]. Think about what this means. After 10–15 years and perhaps $1 billion in costs, you still fail half the time. Although some of the public seem to view the pharma industry as rich with excessive profits, it is a tough industry and requires large investments and remarkable stamina. In 2010, $63 billion was spent on pharma R&D and 22 drugs were approved [7]. This improved somewhat in 2011 with 35 drugs gaining FDA approval [8] and then again in 2012 with 39 new drug approvals [9].

Identification of the active ingredient is not the only thing required to bring a drug to market. In order to optimize the stability, release, and dosing of the active, the delivery method must be studied. The active can be delivered orally, by injection, inhalation, topically such as with a cream, or transdermally as in smoking cessation or seasickness patches. For solid drugs, crystallinity and particle size can have an influence on how the drug is released. Many actives are poorly water soluble. Others are hygroscopic. Others are either moisture or acid sensitive. All of these things need to be considered by the formulation chemist. They play a major role in patient acceptance and compliance and also the therapeutic value of the drug.

For successful drugs, there is a financial reward. Global sales are about $840 billion annually with the top ten products averaging $7.5 billion annually [10].

10.2 REGULATION

The regulation of pharmaceuticals by the United States government can trace its roots back to 1906 when President Theodore Roosevelt signed the Food and Drug Act, also known as the Wiley Act [11]. This act dealt mainly with misbranding. Remember that this was the era of snake-oil salesmen and magical elixirs. The Food, Drug and Insecticide organization, later to be the Food and Drug Administration (FDA) was formed in 1927, but the laws were weak.

In 1937, a chemist and pharmacist at the S.E. Massengill Co. formulated a liquid version of the antibiotic sulfanilamide by using diethylene glycol instead of the more commonly used ethanol. He did this without testing the diethylene glycol for safety [12]. Over one hundred people died, many of them children. Tragically, history repeated itself in 2006 in China when at least five people died from kidney failure after being injected with a gall bladder drug that used diethylene glycol [13] and again in 2007 in Panama when more than 300 people died after taking cough medicine manufactured with diethylene glycol that was believed to be glycerin. The diethylene glycol was manufactured in China and believed to have been relabeled as glycerin by a middleman in Spain [14]. The Massengill tragedy was perhaps the worst, but not the only tragedy of the time due to a lack of regulation. About the same time, a man picked up horsetail weed from a railroad track, added herbs and water and sold it as Banbar, a cure for diabetes [15]. People stopped taking their insulin and died. The outrage of these and other tragedies led to a public outcry and the passage of the Food, Drug and Cosmetic Act of 1938 which mandated a premarket review of the safety of all new drugs [16].

In the late 1950s, thalidomide began to be used in Western Europe for insomnia and to alleviate nausea associated with pregnancy. In 1960, Richardson-Merrell Inc. submitted a new drug application (NDA) to market thalidomide in the U.S. The FDA did not approve the application because the reviewing medical officer, Frances Kelsey, found the proof of safety to be lacking [17]. Later thousands of babies whose mothers had taken thalidomide were born in Europe with horrible deformities. The FDA had prevented a similar widespread tragedy in the U.S. In response to the thalidomide disaster, the U.S. passed the Kefauver–Harris Amendments of 1962, which strengthened the rules for drug safety and required manufacturers to prove their drugs' effectiveness.

The Generic Drug Enforcement Act of 1992 was the result of a scandal involving allegations and proof that various generic drug manufacturers had provided illegal gratuities to FDA reviewers and falsified data. Among other provisions, the statute provided for disbarment of individuals and companies from submitting or assisting in any Abbreviated New Drug Applications (ANDA) [18].

ANDA is the request to obtain approval to market a drug that has come off patent, a generic drug. Information demonstrating that the generic version is bioequivalent to the brand-name drug and does not infringe patents is required. The generic drug must dissolve in about the same time and enter the blood-stream at about the same rate as the brand-name version. The first company to file an ANDA is given 180 days of exclusivity for the generic version. The FDA maintains a list of approved drug products in a book entitled "Approved Drug Products with Therapeutic Equivalence Evaluations." Because of the color of the book, it is commonly referred to as the "Orange Book." The FDA maintains an electronic version of the Orange Book on their website [19]. It is searchable by active ingredient, by proprietary name, and by applicant holder.

As the need arises, Congress continues to pass laws to regulate the industry. The laws are compiled in the Code of Federal Regulations (CFR). The CFR is the codification of the rules published in the Federal Register by the departments and agencies of the Federal Government. It is divided into 50 titles that represent broad areas subject to Federal regulation. Title 21 governs pharmaceuticals as well as many other areas such as food, cosmetics, medical devices, and tobacco products. Title 21 is lengthy; it has nine volumes, three chapters, and more than 1400 parts. Each part deals with a topic. For example, in Chapter One, Part 201 covers labeling. Within Part 201, there are over 60 sections. Some numbers are saved for future use, but they run up to Section 326. Each section is on a specific topic and can have subsections. For example, Subsection 301 of Section 201 is a notice to manufacturers, packers, and distributors of estrogenic hormone preparations. The 21CFR regulations can be searched online at government websites [20, 21].

The rules are administered by the Food and Drug Administration (FDA) which is an agency within the Department of Health and Human Services (HHS). HHS is a cabinet position within the government. There are more than 50,000 employees and the 2014 budget is $975 billion. The FDA has about 12,000 employees to protect the public health by assuring the safety, effectiveness, and security of human and veterinary drugs, vaccines and other biological products, medical devices, our nation's food supply, cosmetics, dietary supplements, and products that give off radiation. Even with such large resources, each year the FDA inspects only about 40% of domestic facilities and 11% of international facilities that supply to the U.S. [22].

In 2012, the Food and Drug Administration Safety and Innovation Act (FDASIA), was signed into law. It expands the FDA's authorities and strengthens the agency's ability. FDASIA gave the FDA a new expedited drug development tool, known as the "breakthrough therapy" designation. The purpose is to expedite the development and review of new drugs with preliminary clinical evidence that indicates the drug may offer a substantial improvement over available therapies for patients with serious or life-threatening diseases.

One way that the FDA protects public health is to make sure that pharmaceuticals are made by good manufacturing practices (GMP). Because this is not static, but is expected to continuously improve, the regulations call for current good manufacturing practices (cGMP). The goal is to protect public health, but the concept is to build quality into the products with the idea that products manufactured by cGMP are safe, properly identified, and of high quality.

There are many aspects of cGMP. They include elements such as organization, documentation, training, corrective actions, raw material qualification, and equipment validation. The organization should have a quality assurance unit responsible for approving or rejecting procedures or specifications, compliance with procedures, and approval or rejection of manufactured product. The organization must have adequate staffing; they must be trained, and the training documented. Equipment must be calibrated and validated. Approved written procedures for production and process controls must be maintained and any deviations properly documented. Production and laboratory records, equipment cleaning logs, and distribution records must be maintained for at least one year after the expiration date of the drug.

Standard Operating Procedures (SOPs) are critical to cGMP compliance. To convince you of this, think about paper airplanes. Everyone thinks they know how to make a paper airplane. It is something we all did in our childhood. However, ask 10 people to make and fly a paper airplane and you will witness 10 varied flights. Some will travel less than a foot; others may soar more than 50 feet. This huge variation is not acceptable in drug manufacture. Imagine if there were a 50-fold variation in purity or potency. Detailed SOPs must be followed to ensure product quality and uniformity. There must be enough detail to make sure that different shifts or different people all make the same product. The details may seem like overkill and it can be tedious to read or write these procedures, but the consequences of a minor change influencing the quality of the product make it absolutely necessary to follow SOPs. Any changes from SOPs must go through a change control process. Documentation of the change control is critical.

When equipment is installed, it has to be qualified to ensure that it has been properly installed and operational qualification performed to make sure it is operating properly. Performance qualification is done to ensure that the equipment operates correctly in the actual operating environment. Traceability is important so raw materials must be controlled and traceable to products. Batch records and equipment use records should show ingredients, equipment, containers, labeling, and personnel. Lot or batch numbering is the key to traceability.

To be cGMP compliant, there must be Good Documentation Practices (GDP). It is important to write things down in permanent indelible black ink, to leave no doubt, and to sign and date the entry. Correction fluid should not be used. If there is an error: draw a line through without obscuring the original entry; make the correction; sign and date the correction.

When the FDA inspects a facility and finds situations of objectionable conditions or practices, at the conclusion of the inspection they issue a Form 483. This represents a list of GMP concerns in the judgment of the inspector. By monitoring 483s, which are public record, companies can maintain currency in GMP. Warnings letters can also be issued by the FDA. They represent concerns not only of an investigator, but of District and/or Center compliance officers. Possible repercussions include product recall, seizure, injunction, monetary fine, debarment, disqualification, license suspension or revocation, prosecution, or denial of access to United States market for foreign suppliers.

Drug development can be divided into different stages. Initially, there is the discovery activity. The discovery activity is not governed by regulatory standard [23]. In the next stage, toxicology and safety pharmacology studies, with a potential extension to pharmokinetics and bioavailability are done. These studies can be termed non-clinical or preclinical because they are not done with humans. The primary purpose of this stage is safety and it requires compliance with Good Laboratory Practices (GLP). Subsequent stages involve clinical studies with humans. Here Good Clinical Practice (GCP) is required. Also at this stage, and through the lifetime of the drug, all manufacturing must be done according to cGMP. In the commercial manufacturing of an approved drug, the quality control labs are governed by cGMP and not by GLP. GLP is described in part 58 of Title 21. This part is divided into 11 subparts with each subpart describing an aspect of GLP.

GLP can be broken down into six requirements [24]:

1. Responsibilities should be defined for the sponsor management, study management, and quality assurance unit.
2. All routine work should follow SOPs.
3. Facilities should be large enough and constructed to ensure the integrity of the study.
4. Test and control articles should have the right quality and instruments should be calibrated and well maintained.
5. People should be trained or otherwise qualified for the job.

6. Data should be acquired, processed and archived to ensure data integrity.

Because the pharmaceutical industry is a global industry, there is a need for regulations beyond the United States. The International Conference on Harmonization of Technical Requirements for Registration of Pharmaceuticals for Human Use (ICH) began in 1990. It is a project that brings together the regulatory authorities of Europe, Japan, and the U.S. and experts from the pharmaceutical industry in the three regions to discuss scientific and technical aspects of product registration. The purpose is to make recommendations on ways to achieve greater harmonization in the interpretation and application of technical guidelines and requirements for product registration in order to reduce or obviate the need to duplicate the testing carried out during the research and development of new medicines. ICH guidelines represent agreed-upon scientific guidance for meeting technical requirements for registration within the three ICH regions – EU, U.S., and Japan. ICH publishes guidelines on a variety of topics related to the pharmaceutical industry. For example, there are documents on stability testing, on validation of analytical procedures, on impurities in new drug substances, and on many other subjects.

10.3 SYNTHETIC CONSIDERATIONS

Most pharmaceuticals are small synthetic molecules, most of which are based upon nitrogen heterocycles. These molecules are typically made by multi-step syntheses. The best synthesis is one that gives the desired compound with the best quality and the lowest cost. Often lowest cost is determined by the number of steps, the yield of each step, and the complexity of each step. It is important for pharmaceutical companies to develop attractive syntheses and considerable resources are expended in process development.

As an example, consider the drug losartan K (Cozaar), which is commonly prescribed for hypertension. The drug was invented by Dupont workers and developed and marketed with Merck & Co. Inc. The patent [25] covering the composition expired in 2009 and the drug is now offered in generic form. Several early syntheses originated from 4′-methyl-biphenyl-2-carbonitrile.

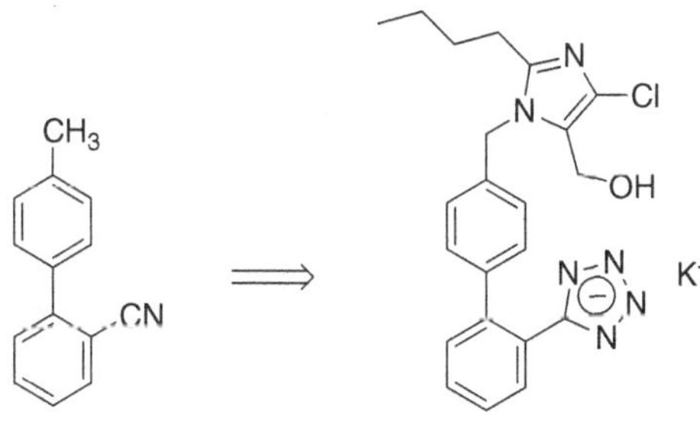

CH_3 CN ⟹ N N Cl OH N N N N K^+

losartan K

Of these, the best synthesis [26] involves four steps and gives an overall yield of 11.4%. The synthesis also has a radical benzylic bromination step using carbon tetrachloride as a solvent. There are industrial hygiene issues with large scale use of carbon tetrachloride because of its health effects. Also with an overall yield of 11.4%, there are many impurities and purification can be a problem. Even without these problems, the starting carbonitrile itself is not readily available, requiring either a low yield (11%) copper coupling reaction or a higher yield synthesis employing a Grignard reagent. Grignard reagents are difficult to handle in large scale because of the flammability issues with the reagent and solvent.

Losartan K is a widely prescribed drug and the early synthetic routes are problematic. Because of this, the inventing company has a motivation to develop improved processes. An improved route can improve profitability and in some instances can be patented, thereby extending the exclusivity for the inventing company. When a drug goes generic, as Losartan K has, the company with the best synthesis will often be in the best competitive position to market the generic compound. For this reason, other companies have a motivation to develop improved syntheses of large volume drugs. They too have an opportunity to obtain patent coverage for new synthetic processes. Because of these reasons new processes have been developed by workers at Dupont-Merck, Abbott, Wyeth-Ayerst (now Pfizer), Roche, Synthelabo, Johnson and Johnson, Zeneca, and Sanofi [27]. That so many companies have worked in this area demonstrates the importance of having the best synthesis.

Losartan K is an example of a drug that can exist in different crystalline forms. The occurrence of different crystalline forms of the same drug is called polymorphism. Polymorphs can have different chemical properties such as dissolution rate, hygroscopicity, stability, and bioavailability. Therefore, it is important for pharmaceutical companies to understand the possibility of polymorphism. Ignoring this could lead to possible quality problems. There have been instances where a specific polymorph has been patented after a patent had issued for the pharmaceutical active itself. This effectively extends the patent protection of a drug.

An attractive synthesis is high yield, has few steps, results in good purity, and avoids conditions or reagents that are problematic on a large scale. A multi-step synthesis, even if all the steps are individually high, can have a low overall yield. For example, a ten-step synthesis with each individual step being 90% yield has an overall yield of 35% (0.90^{10}) and if each step were 80% yield, the overall yield would be about 11%. One low yield step can be a problem. A ten-step synthesis with nine reactions having a 90% yield and one being 10% yield has an overall yield of 3.9%. If faced with a situation like this, it is generally better if the low yield reactions take place early in the synthetic sequence. Otherwise, you have to do large-scale reactions throughout the sequence only to lose yield on a late reaction. If the low yield reaction is early in the sequence, that reaction needs to be a large-scale reaction, but subsequent steps can be done on a smaller scale.

Most syntheses involve sequential steps, but there are many advantages to a convergent synthesis where the different segments are assembled at or near the end of the synthetic process. One example of a convergent synthesis is the synthesis of methotrexate. Methotrexate is used in chemotherapy and acts by inhibiting the formation of folic acid. Methotrexate is also used to treat rheumatoid arthritis. It can be prepared in a one-step convergent (dubbed a shotgun reaction on the production floor) from tetraaminopyrimidine, p-(N-methylamino)-benzoyl glutamic acid and a three-carbon synthon such as dibromopropionaldehyde or 1,1,3-tribromoacetone [28].

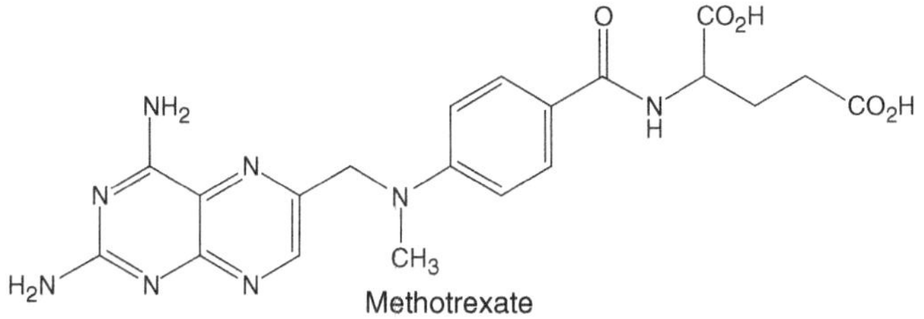

Methotrexate

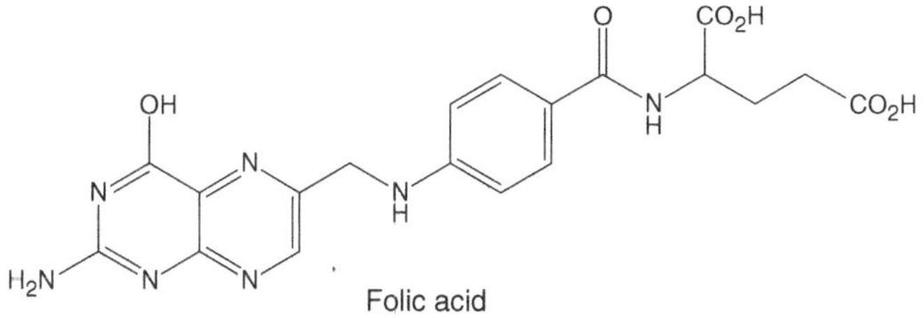

Folic acid

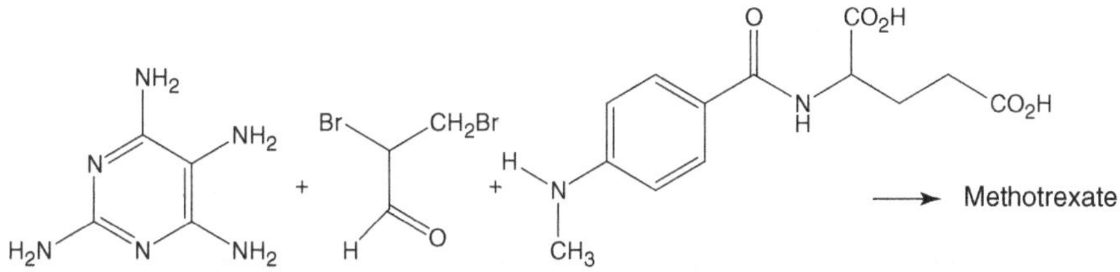

→ Methotrexate

The purity of the final product is a critical factor in selecting a synthetic route. Any impurities must be below 0.1% by weight or must be studied for their health effects. A synthesis which gives impurities which are difficult to remove is often rejected even if the yields are higher. Scale-up considerations are also important. The use of high pressures, flammable solvents, or processes which create a lot of waste all add cost due to equipment, safety, and waste-disposal needs. Diethyl ether can be used safely in the laboratory, but its low boiling point and flammability make it problematic in larger reactions. A process with chromium, mercury, or copper waste, for example, might be rejected because of the difficulties and costs in properly disposing of the waste. Things that a laboratory chemist may not consider, such as how fast something filters, can make a tremendous difference when the reaction is done on a large scale.

10.4 CHIRALITY

The term chiral comes from the Greek word for hand. A chiral carbon has four different groups bonded to the carbon and the mirror image is not superimposable and therefore a different compound. This is true for our hands – our left hand is the mirror image of the right and our hands are not superimposable. Our shoes are another example of nonsuperimposable mirror images. They have the same physical properties such as weight and density. However they can be distinguished by a chiral material – in this example by our eyes or by putting them on our feet which are also chiral. You can only put the left shoe on the left foot. Contrast this with socks which are achiral. Socks are superimposable mirror images and there is no left sock and right sock. They are the same. Lactic acid is an example of a compound with a chiral center.

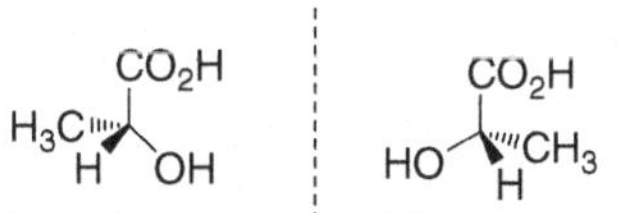

CO_2H H_3C H OH — R-lactic acid

CO_2H HO CH_3 H — S-lactic acid

The mirror image of R-lactic acid is not superimposable and is a different compound. Non-superimposable mirror images are called enantiomers. Enantiomers have the same physical properties. For example, the boiling point, melting point and solubility are the same. They differ in the rotation of plane polarized light. One enantiomer rotates plane polarized light in a clockwise direction; the other enantiomer in a counterclockwise direction. The ability to rotate plane polarized light is called optical activity. A 50–50 mixture of enantiomers is called a racemic mixture or a racemate. Because each enantiomer rotates light in an equal and opposite direction, in a racemic mixture the rotations cancel each other out and there is no net rotation. A racemic mixture is said to be optically inactive. Separation of the enantiomers in a racemic mixture is called resolution. This can be done analytically by HPLC using a chiral stationary phase. If there are x number of chiral centers, there can be up to 2^x stereoisomers. The amino acid, threonine, has two chiral centers and can have up to 2^2 or four stereoisomers.

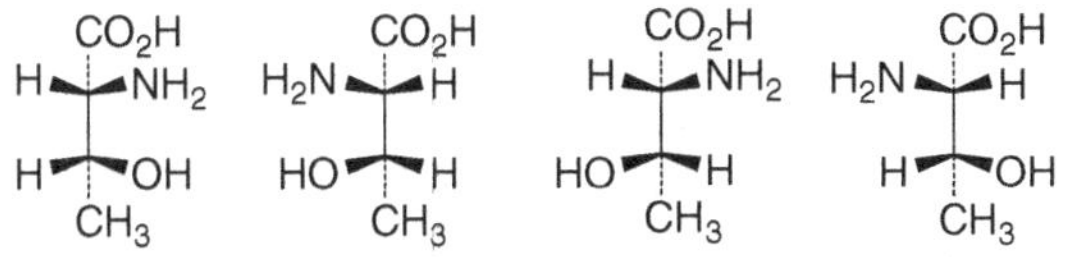

2R,3R 2S,3S 2R,3S 2S,3R

Threonine stereoisomers

The relationship between the 2R,3R and the 2S,3S stereoisomers is that they are nonsuperimposable mirror images of each other; they are enantiomers. Similarly, the 2R,3S and the 2S,3R stereoisomers are enantiomers of each other. The relationship between the 2R,3R and the 2R,3S stereoisomers is that they are diastereomers. Diastereomers are stereoisomers that are not mirror images. Diastereomers generally have different, albeit similar, physical properties. Of the four stereoisomers of threonine, only the 2S,3R stereoisomer occurs naturally and is an essential human nutrient [29].

Consider 2,3-diaminobutane. There are two chiral centers and therefore we could have up to four stereoisomers. The four possibilities are arbitrarily labeled a-d. Stereoisomers a and b are nonsuperimposable mirror images and are enantiomers. Stereoisomers c and d are also mirror images but they are superimposable. Therefore they are the same compound. There are only three stereoisomers, a-c.

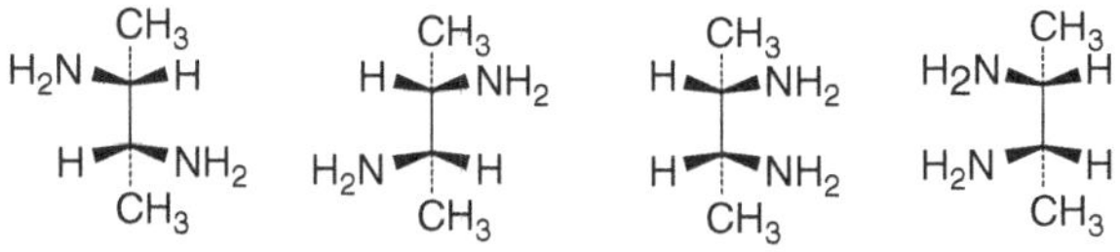

a b c d

2,3-diaminobutane

If we look closely at stereoisomer c, we see that there is a plane of symmetry within the molecule. Stereoisomer c is achiral yet has chirality centers. Compounds like this are called meso compounds. A meso compound is optically inactive. The relationship between a and c (or between b and c) is that they are diastereomers.

Plane of symmetry

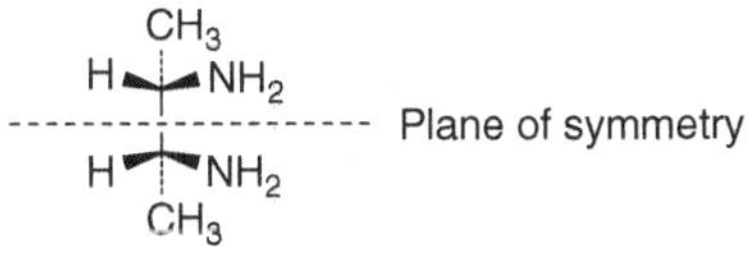

c

The human body can distinguish between enantiomers. For example, consider the enantiomers of limonene. The R-(+) enantiomer has the odor of orange and the S-(−) enantiomer the odor of lemon. The designation (+) indicates that plane polarized light is rotated in the clockwise direction or that the compound is dextrorotatory and (−) indicates counterclockwise rotation, also known as levorotatory. R and S indicate the positioning of the substituents on the chiral carbon. Some R compounds are dextrorotatory and other R compounds are levorotatory. R-(−) carvone has a spearmint odor and S-(+) carvone has a caraway odor. R-(−) amphetamine has a musty odor and S-(+) a fecal odor. There are examples of other compounds where enantiomers do not have different odors, but there are several examples that prove that enantiomers can have different odors [30].

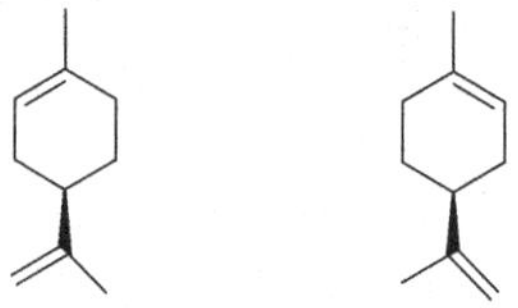

R-(+) limonene S-(−) limonene

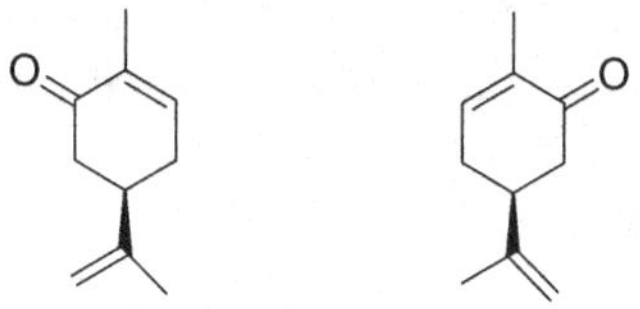

R-(−) carvone S-(+) carvone

S-(+) amphetamine R-(−) amphetamine

It is also true that enantiomers can have different bioactivity, bioavailability, and pharmacokinetics. There are many examples. One is albuterol, used for over 40 years as a racemic mixture to treat asthma. It is the R(−) enantiomer which is the effective bronchodilator. There are some side effects with the use of racemic albuterol. The S(+) enantiomer does not exhibit the beneficial selective binding. At best, it can be considered inert and there is even some indication that it can act as an agonist with effects contrary to the R(−) enantiomer. The S(+) enantiomer is metabolized up to 10 times more slowly and therefore persists in the circulation up to 12 hours [31]. Even if the S(+) enantiomer were inert, a patient taking the racemic mixture needs twice the amount versus taking the effective R(−) enantiomer.

A method of treating asthma with optically pure R(−) enantiomer has been patented [32].

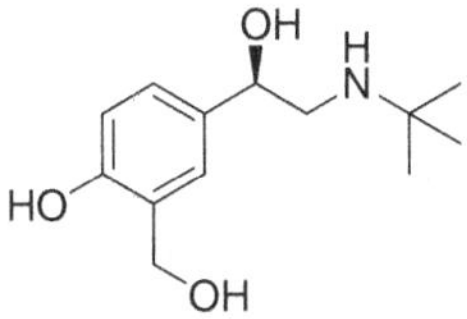

R(–)-albuterol

Another example is propoxyphene [33]. The (2R,3S)-enantiomer, sold as the hydrochloride and once marketed as Darvon®, is an analgesic. The 2S,3R enantiomer is an antitussive and was once marketed as Novrad®. Today, neither is marketed in the United States. Darvon® was removed in 2010 and Novrad® removed about thirty years earlier. The scientists at Lilly who discovered these compounds were clever in their naming. NOVRAD is the mirror image of the name, DARVON.

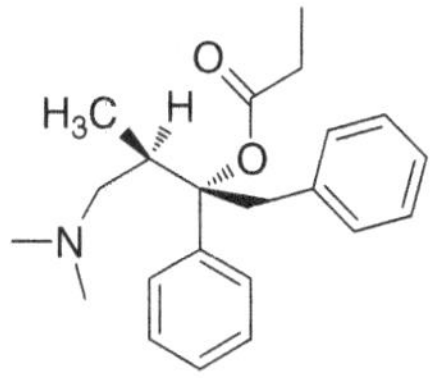

2R,3S-propoxyphene
Darvon

The practice of patenting a single enantiomer in a previously known effective racemate is sometimes called a "chiral switch." Theoretically you could imagine that a pharmaceutical company that invented a new active drug could patent the racemic mixture and obtain exclusivity for 20 years. At the end of the 20 years, they could patent the more effective enantiomer and gain another 20 years for a total of 40 years of exclusivity. However in practice, this does not happen today. A pharmaceutical company that invents a new chiral active will investigate the enantiomers and attempt to obtain patent protection for both the racemate and the effective enantiomer. Otherwise, they risk the possibility that another company might patent the more effective enantiomer and block the first company from practicing the best form of the drug. As an example, the patent for using optically pure R (−) albuterol was not by the original company that discovered albuterol.

The majority of racemic pharmaceuticals have one major bioactive enantiomer. This is called the eutomer. The other inactive or less active enantiomer is called the distomer [34]. A number of cardiovascular drugs such as many of the beta-blockers, calcium channel antagonists, and ACE inhibitors are racemic with one major bioactive enantiomer. A listing [35] of the top ten products by sales shows three to be large molecules such as proteins or antibodies and of the other seven drugs, two are achiral and the other five chiral. One of the five, Seretide®, is a mixture of two components, one of which, fluticasone propionate is present as a single enantiomer. The other component, salmeterol xinafoate, is present as a racemic mixture. That half of the top five drugs are single enantiomers is consistent with another source [36] that states that 68% of drugs under development are chiral and 75% of these are being developed as a single enantiomer, for an overall total of 51% single enantiomers. A subset of these chiral pharmaceuticals is those molecules containing an amine attached to a chiral carbon. One estimate [37] is that up to 45% of all molecules being developed for pharmaceuticals contain an amine attached to a chiral carbon.

There are many synthetic techniques for stereoselective synthesis and because of the importance much research is currently devoted to expanding these techniques. Omeprazole, marketed as Prilosec®, was the first proton pump inhibitor and is used as a treatment for gastric ulcers. It is a racemic mixture. The S enantiomer has better pharmacokinetics and pharmacodynamics than the racemic mixture and therefore higher efficacy in controlling acid secretion [38]. The S enantiomer is called esomeprazole and is marketed as Nexium®. Nexium® has annual sales of $8.4 billion [39]. The chiral center is at the sulfur and the enantiomers can be separated by derivatizing with a R-(−) mandelic acid chiral auxiliary to form two diastereomers. Unlike enantiomers, diastereomers have different physical properties. The diastereomers can be separated by HPLC and then the chiral auxiliary removed to afford each of the omeprazole enantiomers. Alternatively, the sulfide precursor can be asymmetrically oxidized to form esomeprazole in 99.99% enantiomeric excess (ee) [40]. Enantiomeric excess means the excess amount of one enantiomer over the racemic mixture. If something is 80% ee that means that 20% is a racemic mixture and the other 80% is excess of that enantiomer. In this example, for 100 g, there are 90 g of the predominant enantiomer and 10 g of the lesser enantiomer; in other words, 20 g is a racemic mix (10 g of each enantiomer) and 80 g is excess enantiomer.

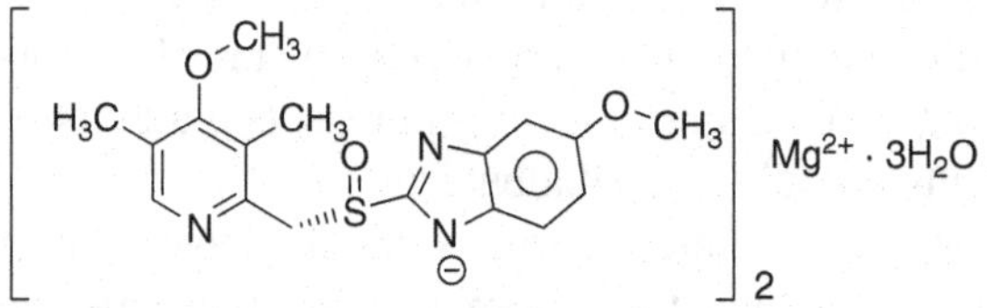

$Mg^{2+} \cdot 3H_2O$

Esomeprazole magnesium

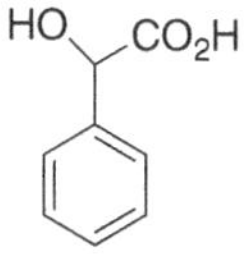

HO CO$_2$H

Mandelic acid

Asymmetric synthesis can also sometimes be done by biocatalysis. This has been used in large-scale syntheses for years. For example, the Reichstein process for the synthesis of vitamin C, established in 1933 and still in use, involves the biotransformation of sorbitol to L-sorbose [41]. L-carnitine is used as a thyroid inhibitor. It is prepared by enzymatic hydroxylation of 4-butyrobetaine [42].

4-butyrobetaine → L-carnitine

Of the methods to prepare single enantiomers, resolution by preparation of the diastereomers from the racemic mixture and then separation of the diastereomers followed by regeneration of the single enantiomer might be the most common. If the racemic mixture has either acidic or basic functionality, this can be accomplished by salt formation. For example, in the synthesis of levothyroxine, used to treat hypothyroidism, the chiral center is resolved at an intermediate stage. The intermediate is a racemic mixture and has a carboxylic acid functionality. When the racemic mixture is treated with a chiral amine to form the ammonium carboxylate, two diastereomers are formed. These can be separated and then the carboxylic acid of the single enantiomer regenerated by treatment with acid.

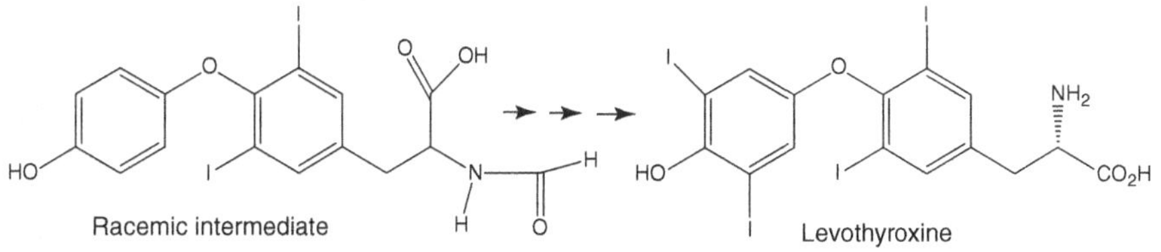

Racemic intermediate → → → Levothyroxine

Asymmetric synthesis is also common. Another technique is kinetic resolution. Kinetic resolution relies upon a difference in reactivity between the two enantiomers. For example this technique can be used in the synthesis of duloxetine. Duloxetine is marketed as the hydrochloride salt under the tradename Cymbalta® as an antidepressant. It is a selective serotonin and norepinephrine reuptake inhibitor. It is the S enantiomer which is used.

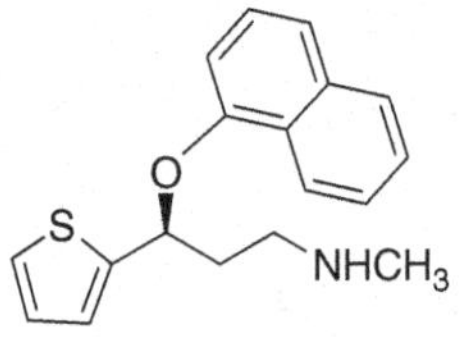

Duloxetine

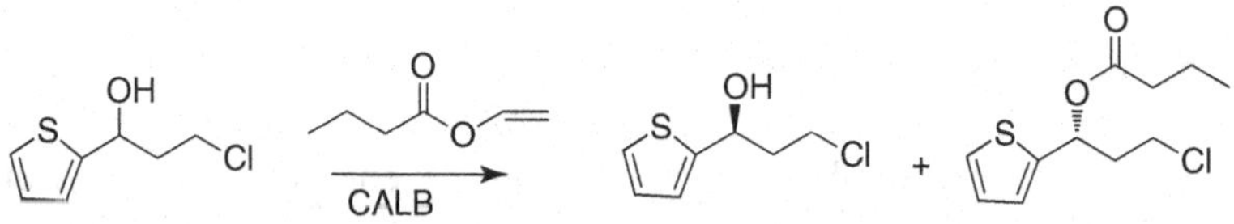

Racemic alcohol

CALB

The precursor alcohol is formed as a racemate and then reacted with vinyl butanoate to 50% conversion in the presence of catalytic *Canidida antarctica* Lipase B (CALB). The undesired R alcohol esterifies faster. After 50% conversion, it has formed the ester and the desired S alcohol remains [43] for subsequent conversion to duloxetine.

One method that avoids the need for asymmetric synthesis or separation of the racemic mixture is to begin with an enantiomerically pure starting material and maintain the chirality throughout the subsequent synthetic steps. This is sometimes called a chiral pool synthesis and is especially useful when the chiral starting material is an inexpensive naturally occurring chemical.

Some drugs can undergo in-vivo chiral inversion. For example, only the S enantiomer of ibuprofen is active. It is over 100-fold more potent as an inhibitor of cyclooxygenase I than the R enantiomer of ibuprofen [44]. However in the body, the R enantiomer undergoes chiral inversion into the active S enantiomer. In the case of ibuprofen, the S enantiomer is not converted to the R. The inversion is unidirectional and a single enantiomer is formed. However, with other chiral actives, in-vivo racemization can occur. One example is thalidomide. This drug was prescribed in Europe for pregnant women as a sleep aid and to counter morning sickness. Its use resulted in many horrible birth defects in the 1960s. Later studies suggested that the R-enantiomer had the desired therapeutic benefit and that the S-enantiomer was mutagenic [45]. However, the R-enantiomer is not suitable for this application because it racemizes in-vivo to form the undesirable S-enantiomer. Racemic thalidomide is used for other therapies. It is the drug of choice to treat leprosy, albeit under conditions carefully controlled by physicians. It has also been approved for the treatment of multiple myeloma, a form of cancer [46].

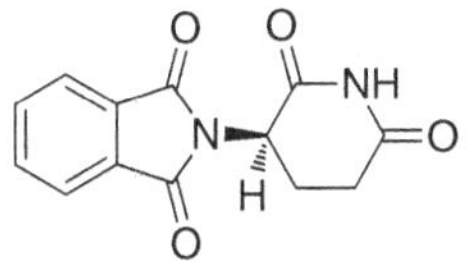

R-thalidomide

Although there have been many advances and discoveries, there remains a need for new pharmaceuticals. Diseases such as cancer, aids, and Alzheimer's claim many victims each year. Despite the long process from discovery to market, the high costs associated with research and clinical trials, and the low probability of success for each lead compound, there is reason to hope that the pharmaceutical industry will make strides in these and other areas in the years to come.

QUESTIONS

1. A seven-step synthesis of a drug has each of reactions one through six having a yield of 95.0% (i.e. 95.0% for reaction one, 95.0% for reaction two, etc.) and reaction seven has a yield of 90.0%. What is the overall yield (use three significant figures)?

2. What is a typical length of time to bring a drug to market (discovery through commercialization)?

a. 6–8 months

b. 2 years

c. 12 years

d. 22 years

3. What are the odds of a compound in the discovery phase being commercialized as a drug?

a. about 1 in 2 (half) make it to commercialization

b. about 1 in 5

c. about 1 in 100

d. about 1 in 250

e. less than 1 in 10,000

4. In developing the most effective cost position for a synthesis, three major considerations are: *(fill in consideration #3 using fewer than 10 words in your answer)*

1. as few steps as possible

2. high yield reactions
3. ____________________________________

5. For a racemic pharmaceutical, typically …
   a. each enantiomer has the same bioactivity
   b. one enantiomer has bioactivity and the other has less activity or is inactive
   c. one enantiomer has bioactivity and the other has the opposite effect
   d. it is impossible to distinguish the bioactivity of the enantiomers

6. What does the c in cGMP stand for?

7. Drug regulations have been strengthened over the years because of various tragedies. Which of the following events did not occur?
   a. children were fed cough syrup with ethylene glycol solvent
   b. a drug to alleviate nausea in pregnant women was used in Europe and caused birth defects
   c. people died after taking horsetail weed as a cure for diabetes
   d. people died from the dye after drinking green beer colored with Green No. 3 dye
   e. a generic drug company falsified data submitted to the FDA

8. (True or False) Detailed SOPs must be followed to ensure product quality and uniformity.

9. A drug entering phase I has about a ___% success rate; typically from the first discovery it takes about _____ years and $_______ to bring a drug to market.
   a. 1% 2 years $50,000,000
   b. 25% 2 years $200,000,000
   c. 85% 12 years $1,200,000,000
   d. 12% 12 years $100,000,000
   e. 12% 12 years $1,200,000,000

10. What is the overall yield of a three-step synthesis that has a first reaction having a yield of 70%, a second reaction having a yield of 80% and a third reaction having a yield of 90%?
   a. 80%
   b. 70%
   c. 60%
   d. 50%

11. Label as True or False: If you are operating under cGMP and you accidentally recorded a value of 23% for an analysis, when the correct value was 32%, it is appropriate to erase the 23% and insert 32%.

REFERENCES

1. John LaMattina. Drug Truths. New York: John Wiley & Sons; 2009:23.
2. John LaMattina. Drug Truths. New York: John Wiley & Sons; 2009:18.
3. Pharmaceutical Research and Manufacturers of America, Pharmaceutical Industry Profile (2010) PhRMA, Washington, DC.
4. John A Vernon, Joseph H Golec, Joseph A Dimasi. Health Econ. 2010; 19:1002–1005.
5. John LaMattina. Drug Truths. New York: John Wiley & Sons; 2009:30.
6. Alex Dmitrienko, Christy Chuang-Stein, Ralph B D'Agostino. *Pharmaceutical Statistics using SAS: a Practical Guide*. SAS Publishing, Cary, NY; 2007:4.
7. Lisa Jarvis. Chemical and Engineering News 2011; 89(9):36.
8. Rick Mullin. Chemical and Engineering News 2011; 89(49):12.
9. Lisa Jarvis. Chemical and Engineering News 2013; 91(5):15.
10. Rick Mullin. Chemical and Engineering News 2010; 88(49):14,18.
11. M Allport-Settle. *Current Good Manufacturing Practices*. PharmaLogika, Inc, Willow Springs, NC; 2009:6.
12. W Steven Pray. *A History of Nonprescription Product Regulation*. Psychology Press, Binghamton, NY; 2003:114.
13. Jean-Francois Tremblay. Chemical and Engineering News 2006; 84(21):11.
14. Rick Mullin. Chemical and Engineering News 2011; 89(20):11.
15. W Steven Pray. *A History of Nonprescription Product Regulation*. Psychology Press, Binghamton, NY; 2003:119.
16. M Allport-Settle. *Current Good Manufacturing Practices*. PharmaLogika, Inc, Willow Springs, NC; 2009:7.
17. Maureen Rouhi. Thalidomide. Chemical and Engineering News 2005; 83(25).
18. DO Beers. *Generic and Innovator Drugs: a Guide to FDA Approval Requirements*. Aspen Publishers Online; 2004: 8–3, 8–4.
19. www.fda.gov.
20. http://www.accessdata.fda.gov/scripts/cdrh/cfdocs/cfcfr/cfrsearch.cfm.
21. http://www.gpoaccess.gov/cfr.
22. Rick Mullin. Chemical and Engineering News 2011; 89(20):12.
23. Mark G Slomiany. *The Indispensable Guide to Good Laboratory Practice*. 2nd Ed. Pinehurst Press, New York, NY; 2009:7.
24. Mark G Slomiany. *The Indispensable Guide to Good Laboratory Practice*. 2nd Ed. Pinehurst Press, New York, NY; 2009:6.

25. DJ Carini, JJV Duncia, PCB Wong. U.S. Pat. No. 5,138,069. 1992.
26. W Cabri, R DiFabio. *From Bench to Market*. Oxford University Press; 2000:125.
27. W Cabri, R DiFabio. *From Bench to Market*. Oxford University Press; 2000:127–142.
28. B Singh, FC Schaefer. U.S. Pat. No. 4,374,987. 1983.
29. John McMurry. *Organic Chemistry*. 7th ed. Brooks/Cole, Belmont, CA; 2008:303.
30. Lester Friedman, John Miller. Science 1971; 172:1044–1046.
31. Editorial. Am. J. Respir. Crit. Care Med. 2006; 174:965–974.
32. Timothy Barberich, James Young. U.S. Pat. No. 5,362,755. 1994.
33. Rick Ng. *Drugs From Discovery to Approval*. 2nd ed. New York: Wiley-Blackwell; 2009:339.
34. L Nguyen, H He, C Pham-Huy. International Journal of Biomedical Science 2006; 2:85–100.
35. Rick Mullin. Chemical and Engineering News 2010; 88(49):14–21.
36. Fulvio Gualtieri. New Trends in Synthetic Medicinal Chemistry. New York: Wiley-VCH; 2000:98.
37. Stephen Ritter. Chemical and Engineering News 2013; 91(10):34.
38. Jie-Jack Li, Douglas Johnson, Drago Sliskovic, Bruce Roth. *Contemporary Drug Synthesis*. New York: Wiley; 2004:22–23.
39. Rick Mullin. Chemical and Engineering News 2010; 88(49):14–21.
40. Jie-Jack Li, Douglas Johnson, Drago Sliskovic, Bruce Roth. *Contemporary Drug Synthesis*. New York: Wiley; 2004:25.
41. Fulvio Gualtieri. New Trends in Synthetic Medicinal Chemistry. New York: Wiley-VCH; 2000:205.
42. Fulvio Gualtieri. New Trends in Synthetic Medicinal Chemistry. New York: Wiley-VCH; 2000:211.
43. Douglas Johnson, Jie-Jack Li. The Art of Drug Synthesis. New York: Wiley; 2007:210–211.
44. L Nguyen, H He, C Pham-Huy. International Journal of Biomedical Science 2006; 2:85–100.
45. S Mohan, et al. International Journal of Pharmaceutical Sciences and Nanotechnology 2009; 1(4):309–316.
46. John LaMattina. Drug Truths. New York: John Wiley & Sons; 2009:113.

CHAPTER 11

Pharmaceuticals–Some Important Drugs

11.1 INTRODUCTION

The FDA approves an average of just over twenty new drugs (New Molecular Entities (NME) per year; in fiscal year 2011 there were 35 new drugs approved. Today there are several thousand approved drugs. There are hundreds of drug targets and many diseases require different and highly specific medications. Most of the drugs are small molecules and most of these are heterocycles [1]. The synthesis of different drugs is complex, often requiring stereoselectivity and has been the subject of several books [2–4]. A comprehensive discussion of this topic would require several volumes and is beyond the scope of this text.

Although most drugs are based upon small molecules, there is an increasing number of large molecule biologics. For example, adalimubab (Humira®, Abbott), the number one biologic therapy, is made from recombinant DNA technology and has over 1,000 amino acids and a molecular weight of approximately 148 kd [5]. T lymphocytes (T cells) are a type of white blood cell that release cytokines that promote the destruction of tissues surrounding the joints. One cytokine is tumor necrosis factor α (TNFα) and adalimubab is a TNF blocker. Because of this, it is prescribed for rheumatoid arthritis and several other indications. In 2011, 585,000 patients were treated with Humira and sales were $7.9 billion [6]. Other biologics that are also TNF blockers include etanercept (Enbrel®, Amgen Inc.), infliximab (Remicade®, Janssen Biotech, Inc.), certolizumab pegol (Cimzia®) and golimumab (Symponi®, Janssen Biotech, Inc.).

To give the reader the general flavor of the topic of pharmaceuticals, several classes of drugs are illustrated. The groups were chosen because of their importance in preventing disease and for the number of people that need the

Fundamentals of Industrial Chemistry: Pharmaceuticals, Polymers, and Business, First Edition. John A. Tyrell.

medications. Many people have high cholesterol and cholesterol medication is near the top of many lists of the top drugs. Similarly, hypertension afflicts about one third of the population and often requires more than one type of medication. Proton pump inhibitors are used for the treatment of ulcers, heartburn and other symptoms of gastroesophageal reflux disease which afflicts about 10–20% of the population. An increasing number of people are developing diabetes, the seventh leading cause of death in the U.S. and there are several medications prescribed for diabetics. The last area chosen, antidepressants, affects about 10% of the population.

11.2 CHOLESTEROL DRUGS

Cardiovascular disease accounts for more deaths, about 30% globally, than any other disease and the over-accumulation of cholesterol is known to play a central role in cardiovascular disease [7]. In the United States population older than 19, 15% have cholesterol levels higher than 240 mg/dL [8].

3-Hydroxy-3-methylglutaryl coenzyme A (HMG-CoA reductase) converts 3-hydroxy-3-methylglutaryl coenzyme A to mevalonate, a precursor of cholesterol. Human 3-hydroxy-3-methylglutaryl coenzyme A, also abbreviated as HMGR consists of a polypeptide chain of 888 amino acids [9]. Acetyl-CoA condenses with acetoacetyl-CoA to form HMG-CoA reductase.

Acetyl CoA + Acetoacetyl CoA $\xrightarrow{H_2O}$ HMG-CoA + HS-CoA

Hydride transfer from NADPH (the reduced form of nicotinamide adenine dinucleotide phosphate) gives initially mevaldyl-CoA which then forms mevaldehyde. Another reduction with NADPH gives mevalonate.

HMG-CoA $\xrightarrow{NADPH}$ Mevaldyl-CoA $\xrightarrow{-\text{CoA-SH}}$

Mevaldehyde $\xrightarrow{NADPH}$ Mevalonate

Mevalonate is converted in a series of biosynthetic steps to squalene which in turn is converted to cholesterol.

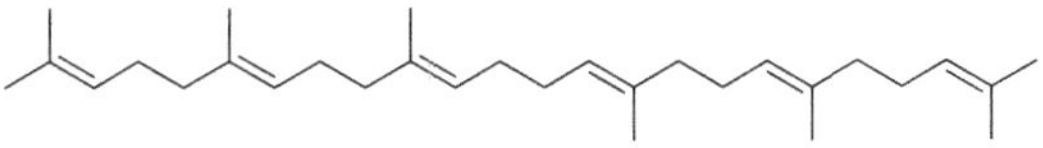

Squalene

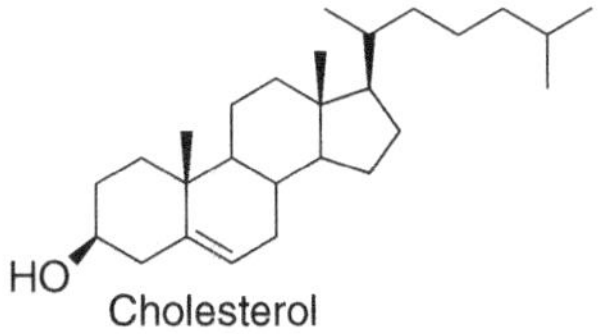

Cholesterol

The formation of mevalonate is the rate-limiting step in cholesterol biosynthesis. Drugs that inhibit HMG-CoA reductase are termed statins and are one of the most widely used types of prescription drugs. All statins share a structural component that is similar to the 3-hydroxy-3-methylglutaryl portion of HMG-CoA reductase. All statins differ from HMG-CoA in being more bulky and more hydrophobic [10]. Statins occupy the binding site of HMG-CoA thus blocking access of this substrate [11]. All statins share this common mode of action but differ in their overall structural, biochemical, thermodynamic, and pharmokinetic properties which influences their efficacy [12].

The first statins were launched by Merck with lovastatin (Mevacor®) in 1987 and then simvistatin (Zocor®) in 1991. Bristol-Myers launched pravastatin (Pravacol®) in 1991; Pfizer commercialized atorvastatin calcium (Lipitor®) in 1997. As measured by annual sales, Lipitor is the most successful drug in the history of the pharmaceutical industry [13]. Rosuvastatin calcium (Crestor®) is AstraZeneca's top drug with $6.6 billion in annual sales [14].

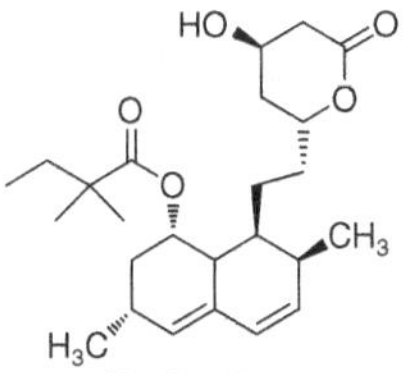

Simvistatin

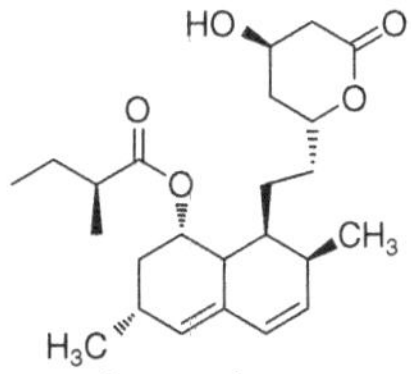

Lovastatin

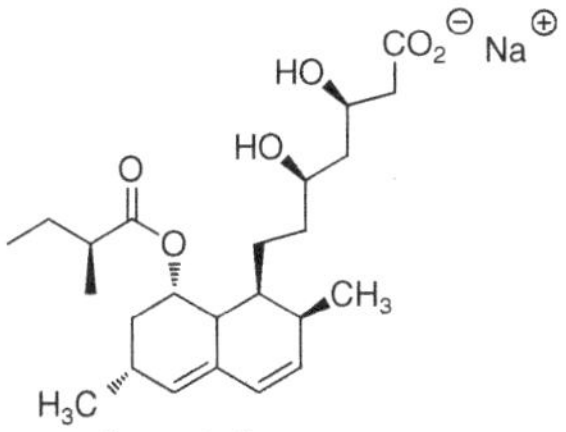

Pravastatin

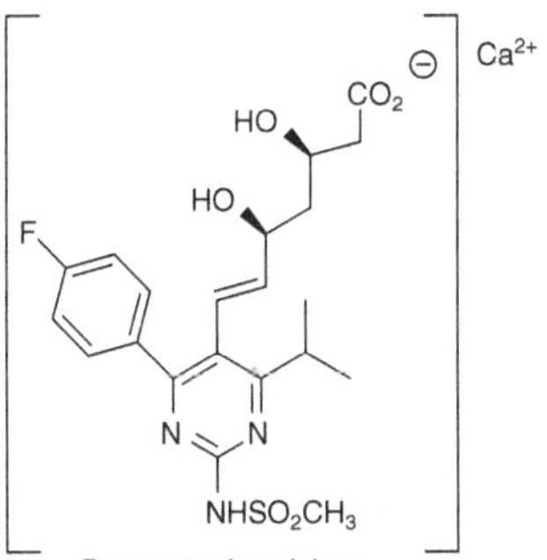

Rosuvastatin calcium

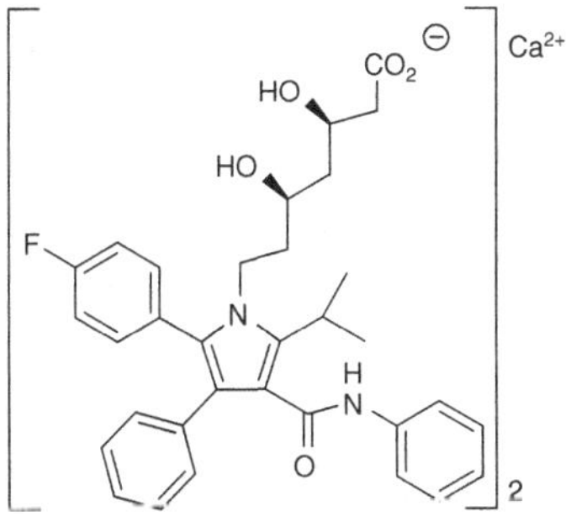

Atorvastatin calcium

Lovastatin can be prepared by a fermentation process[15] in the presence of a specific microorganism. Lovastatin can be converted to simvistatin[16]. Hydrolysis of the ester followed by reclosure of the lactone gives the diol. The less-hindered alcohol can be selectively protected using the bulky t-butyldimethylchlorosilane. The free alcohol can be esterified by the acid chloride in the presences of dimethylaminopyridine acylation catalyst. The silyl ether can be selectively removed by treatment with tetrabutyl ammonium fluoride. The fluoride anion reacts at the silicon without hydrolyzing the lactone or ester.

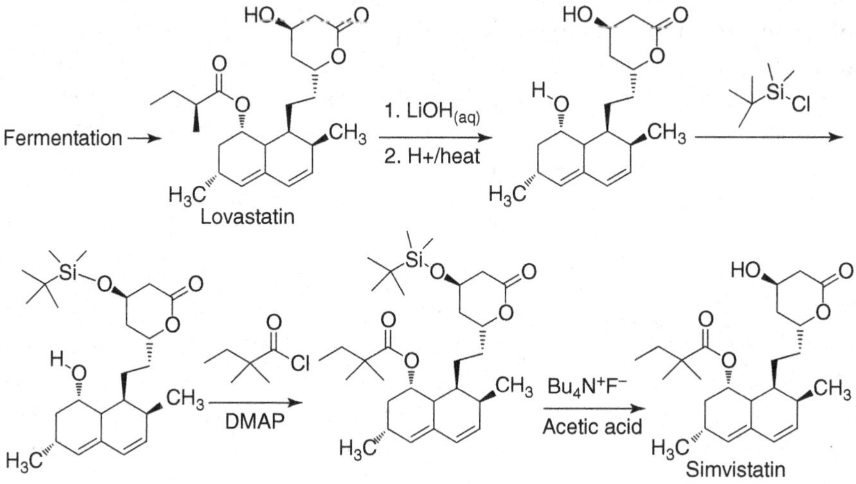

Another treatment for cholesterol is niacin. The use of niacin predates the statins. Niacin is also known as nicotinic acid or vitamin B_3. The name, niacin comes from nicotinic acid and vitamin and was coined to avoid confusion and so that people would not think that the vitamin contained nicotine or that tobacco products contained vitamins. Niacin inhibits lipoprotein synthesis by preventing the secretion of very low density lipoprotein from the liver. Very low density lipoprotein is a precursor of low density lipoproteins (LDL). However there are several adverse side effects with niacin including flushing, warm skin, itching rash, constipation, nausea, heartburn, and problems with liver function. Because of these side effects, niacin is often used in a controlled release form [17] and even in this form is unsuitable for many patients.

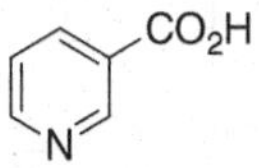

A newer type of cholesterol drugs is based upon the direct inhibition of the uptake of free cholesterol from the small intestine. Perhaps the most prominent in this class is ezetimibe (Zetia®, Schering-Plough Corporation). Note that there are three chiral centers and therefore 2^3 or 8 possible stereoisomers. The stereoisomers exhibit different cholesterol absorption inhibition and therefore a single stereoisomer, as depicted, is administered [18].

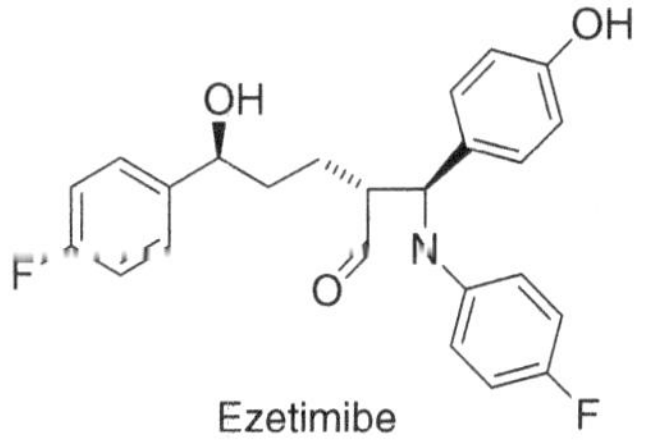

Ezetimibe

11.3 HYPERTENSION

Heart disease is the leading cause of death in the U.S., representing about 25% of all deaths [19]. On average, an American dies every 39 seconds from cardiovascular disease [20]. One third of the United States population older than 19 has hypertension [21]. Hypertension is a major risk factor for cardiovascular disease. Monotherapy adequately controls hypertension in only about 50% of patients and therefore many require a combination of at least two drugs to control blood pressure [22]. Common classes of medications for the treatment of hypertension are angiotensin II receptor antagonists, ACE inhibitors, calcium channel blockers, β-blockers, and diuretics.

The renin angiotensin system is one of the most powerful regulators of blood pressure. Renin, an enzyme, is secreted by the kidney in response to a reduction in blood flow, blood pressure or sodium concentration. Renin then converts angiotensinogen to angiotensin I, a decapeptide. Angiotensin I is cleaved by angiotensin-converting enzyme (ACE) to angiotensin II, an octapeptide which in turn activates angiotensin II receptors. Inhibition of the action of angiotensin II at the receptors prevents the increase in blood pressure caused by this hormone–receptor interaction. Medications that control blood pressure by inhibiting the action of angiotensin II at the receptors are in a class termed angiotensin II receptor blockers (ARBs).

The first receptor antagonist was losartan potassium (Cozaar®, Merck) [23]. Losartan potassium is a prodrug and is converted in the liver to an active metabolite [24]. The term "prodrug" is used for an inactive substance which is metabolized in-vivo to an active form. In other words, it is a drug precursor.

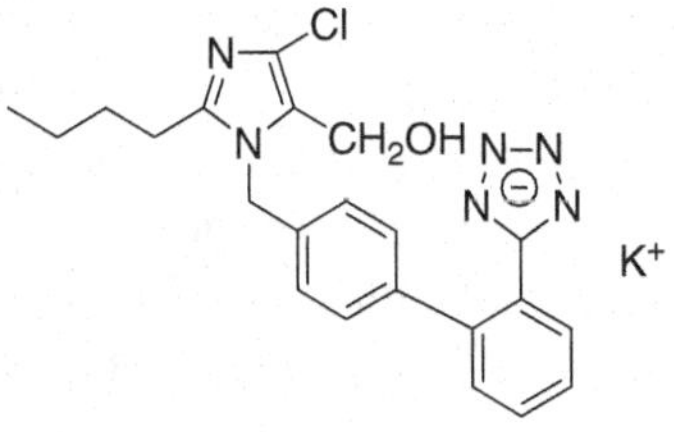

Losartan

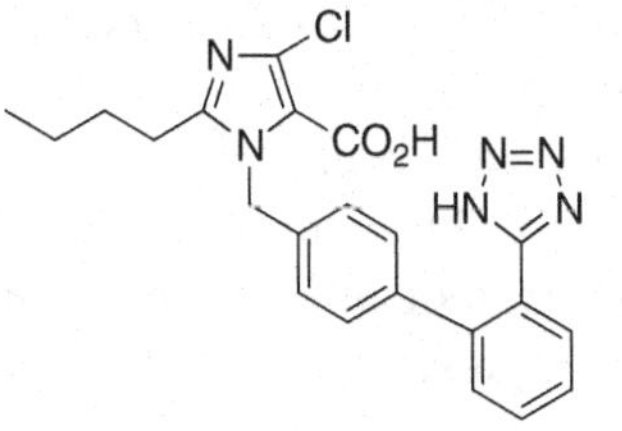

Active metabolite of losartan

The group name sartan was coined and other sartans were developed. Several, such as valsartan (Diovan®, Novartis) and irbesartan (Avapro®, Bristol-Myers Squibb) retained the biphenyl substituted tetrazole ring of losartan.

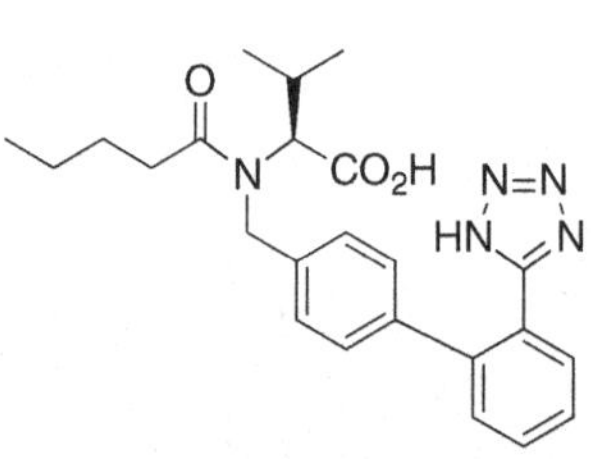

Valsartan

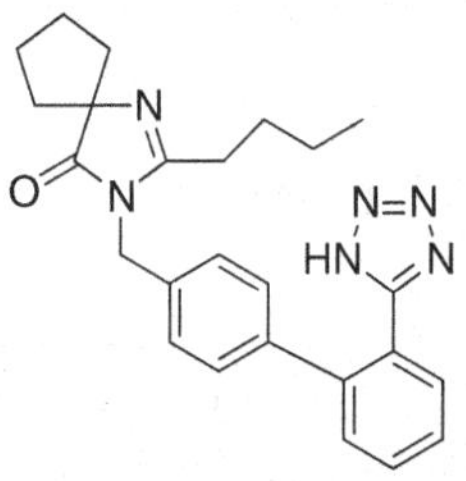

Irbesartan

The synthesis of losartan has been the subject of considerable research. The various syntheses have been discussed [25] in regards to their patent position and their process friendliness or ability to scale. In an early synthesis, the tetrazole is formed by reacting the substituted benzonitrile with sodium azide in a 1,3-dipolar cycloaddition. By protecting the tetrazole as the trityl derivative, the product is soluble in toluene and unreacted azide can be removed by washing with aqueous base. The use of a bulky trityl group also alleviated concerns with exothermic decomposition of the unprotected tetrazole. Benzylic bromination under radical conditions with N-bromosuccinimide was followed by displacement of the bromide with the substituted imidazole. The aldehyde is reduced to the alcohol, the trityl group cleaved under acidic conditions and then treatment with KOH gives losartan.

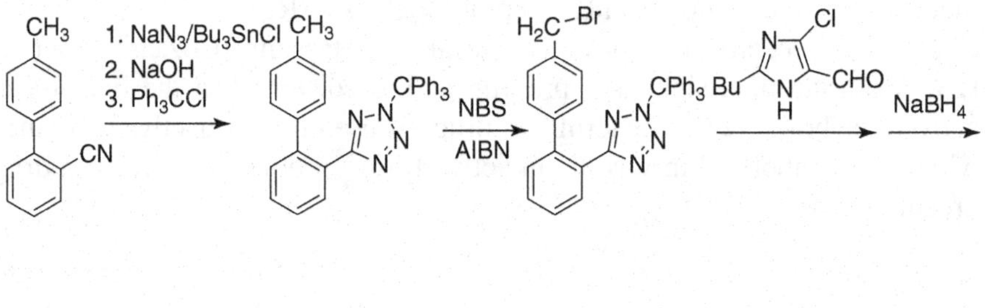

1. NaN_3/Bu_3SnCl
2. NaOH
3. Ph_3CCl

NBS

AIBN

$NaBH_4$

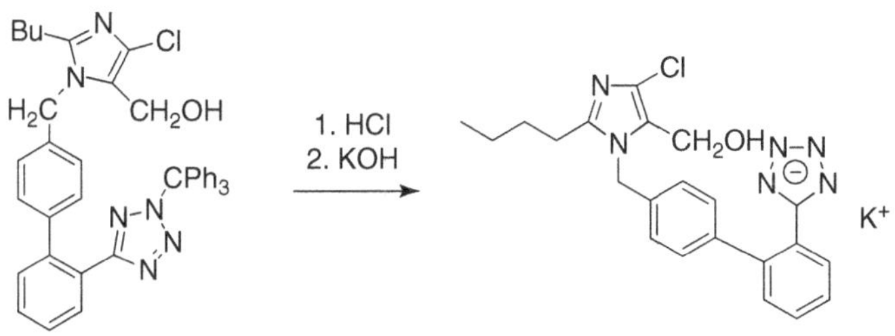

1. HCl
2. KOH

Losartan

An improved and convergent synthesis of losartan depends upon a Suzuki-coupling reaction.

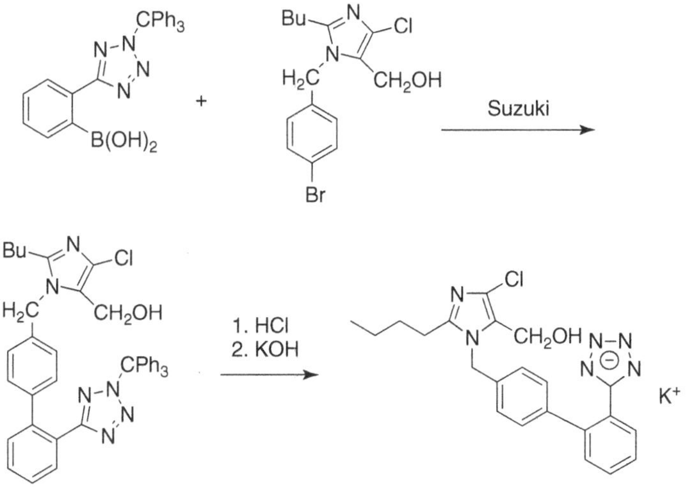

Suzuki

1. HCl
2. KOH

Losartan

Inhibition of ACE blocks the formation of angiotensin II which therefore blocks the renin angiotensin system [26]. Medications that operate by this principle are in the class termed, ACE inhibitors. ACE is a zinc-containing enzyme and the zinc ion catalyzes peptide cleavage from angiotensin I to angiotensin II. ACE inhibitors bind to the zinc cation and therefore inhibit the cleavage. The first ACE inhibitor was captopril (Capoten®, Par Pharmaceutical Companies, Inc), a thiol. The discovery of captopril came from an observation made in the 1960s that an extract of the venom of the Brazilian viper was a potent ACE inhibitor [27]. The extract was a peptide, not suitable for oral administration. To find an orally active form, over the next decade, Squibb spent millions of dollars for research and clinical studies which led to FDA approval of captopril in the early 1980s. Other thiol-containing ACE inhibitors are zofenopril and omapatrilat [28].

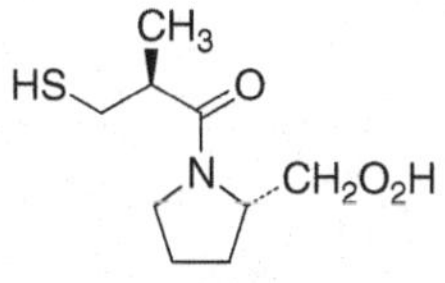

Captopril

The second and largest class of ACE inhibitors replaced the thiol linkage and the potentially related side effects caused by the thiol group by using a carboxylic acid to bind to the zinc. Lisinopril (Zestril®, AstraZeneca) is an example of this type of ACE inhibitor. In some cases an ester prodrug is used which hydrolyzes in vivo to the carboxylic acid. Enalapril and quinapril are examples of ACE inhibitor prodrugs.

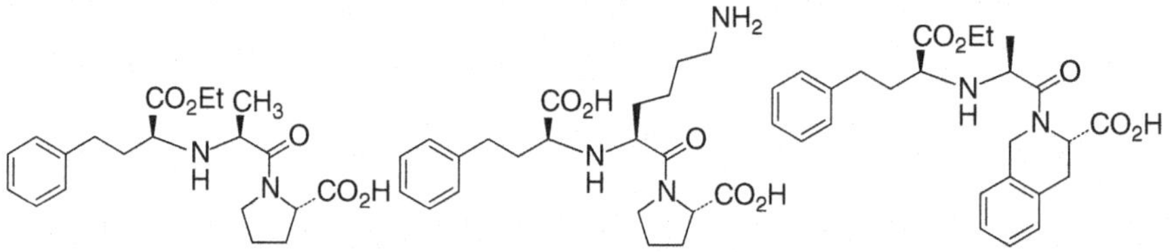

Enalapril Lisinopril Quinapril

Lisinopril can be prepared [29] from the trifluoroacetamide derivative of lysine. Reaction with phosgene gives the N-carboxyanhydride which reacts with proline. Reductive condensation of the amine is followed by hydrolysis of the trifluoracetamide protective group to give lisinopril.

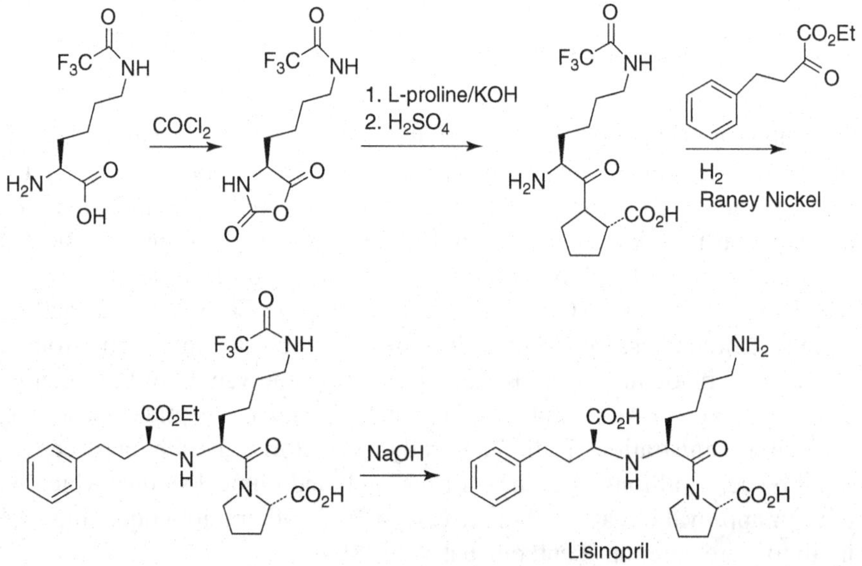

$COCl_2$

1. L-proline/KOH
2. H_2SO_4

H_2
Raney Nickel

NaOH

Lisinopril

A third class of ACE inhibitors is represented by fosinopril sodium (Monopril®, Bristol-Myers Squibb), a phosphinate ester which hydrolyzes in vivo to the phosphinic acid.

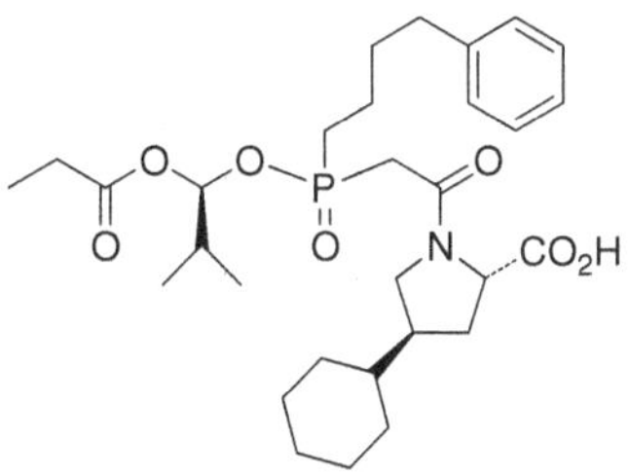

Fosinopril

Calcium channel blockers, also known as calcium antagonists, are a class of hypertension drugs that inhibit the influx of calcium ions through the cell membrane. A decrease in calcium ions results in less contraction of the cardiac and vascular muscles. There is an increase in the diameter of the arteries. This vasodilatation results in a lowering of the blood pressure. Despite their name, calcium channel blockers do not "plug the hole" and physically block the calcium ion channel. Rather, they bind to specific receptor sites [30]. Examples of calcium channel blockers are nifedipine (Procardia®, Pfizer), nicardipine hydrochloride, amlodipine besylate sulfonate (Norvasc®, Pfizer) and verapamil hydrochloride (Calan®, Pfizer). Verapamil has a chiral carbon but is administered as a racemic mixture.

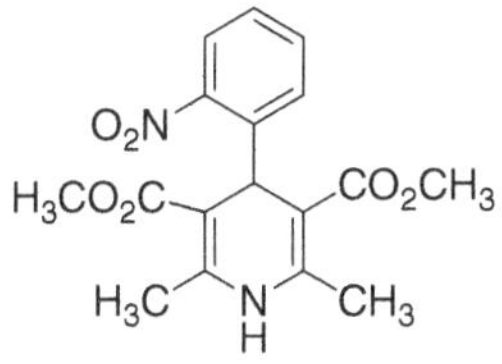

Nifedipine

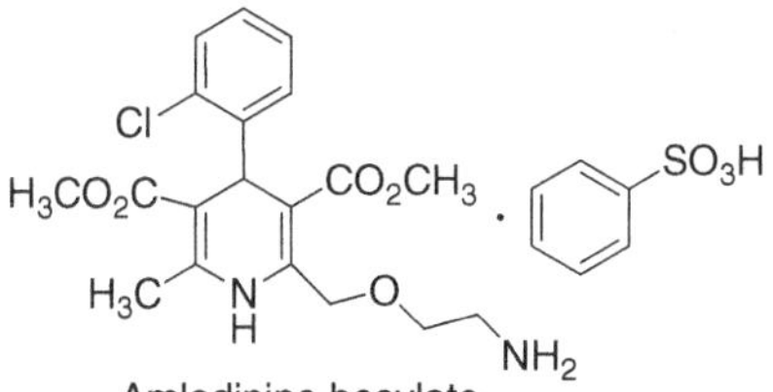

Amlodipine besylate

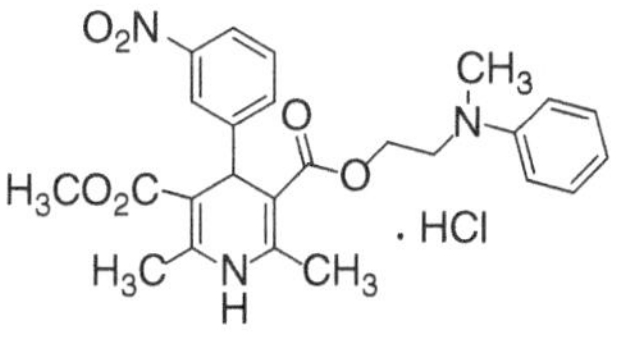

Nicardipine hydrochloride

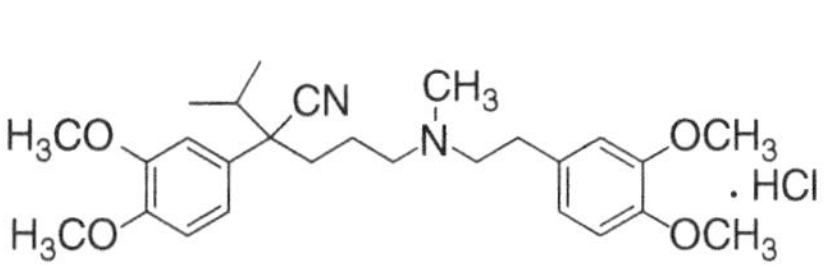

Verapamil hydrochloride

Nifedipine can be made by a Hantzsch synthesis in a one-step convergent synthesis. However, the product purity is problematic and it is better to synthesize nifedipine in a series of steps [31].

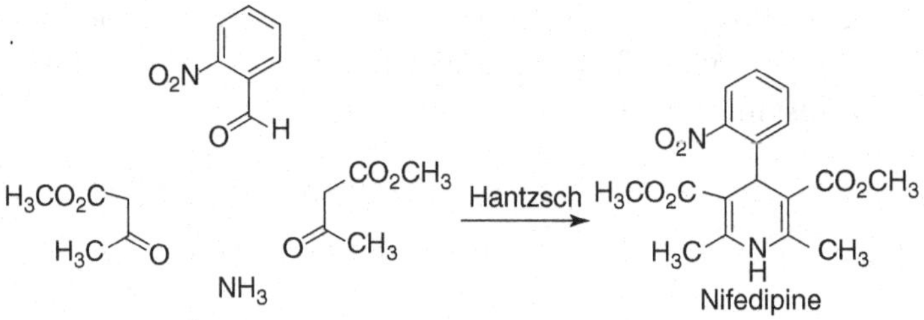

Hantzsch

Nifedipine

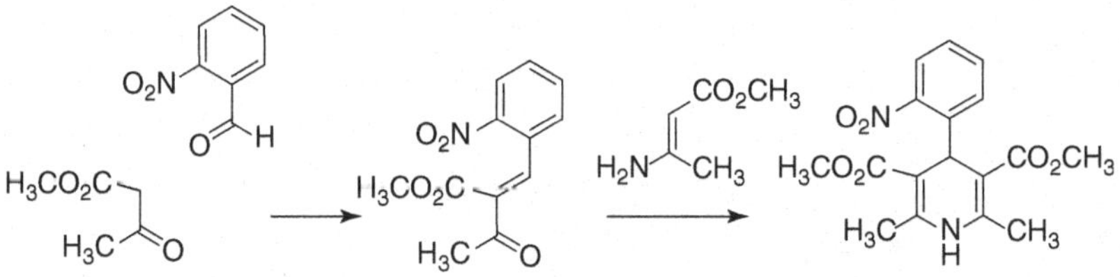

Nifedipine

Beta blockers represent another class of drugs used for hypertension. They block beta-adrenergic receptors, preventing adrenaline (epinephrine) from stimulating these receptors. In this manner, they slow the heart rate and reduce the force with which the heart muscle contracts, thereby lowering blood pressure. Examples include atenolol (Tenormin®, AstraZeneca), carvedilol (Coreg®, Glaxo SmithKline), and metoprolol, sold as the tartrate salt (Lopressor®, Novartis) or the succinate salt (Toprol®, AstraZeneca). For each of these compounds, the carbon bearing the hydroxyl group is chiral and they are all administered as racemic mixtures.

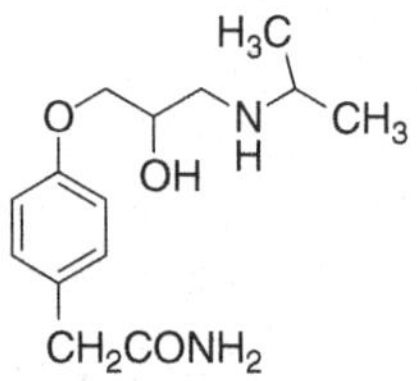

Atenolol

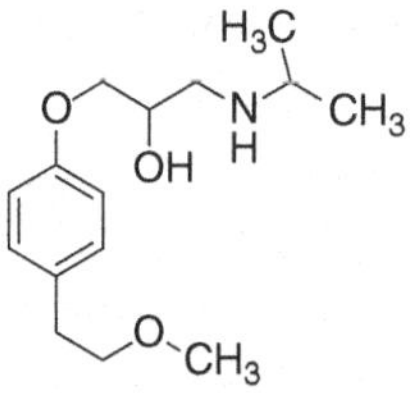

Metopropol

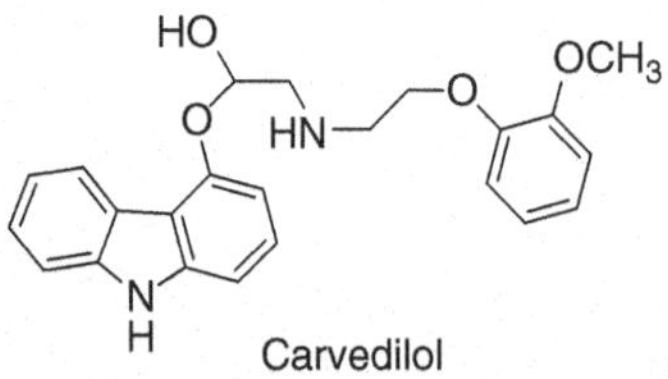

Carvedilol

Diuretics increase the rate of urine formation. One common class of diuretics is the thiazide type represented by hydrochlorothiazide. The thiazides affect the renal tubular mechanisms of electrolyte reabsorption and therefore directly increase excretion of sodium and chloride ions. Hypertension has been linked to high sodium ion content and diuretics are used to combat hypertension by decreasing sodium levels. The thiazides also can increase urinary potassium ion loss. Sometimes they are coadministered with other drugs to compensate for potassium loss. One example is irbesartan-hydrochlorothiazide (Avalide®, Bristol-Myers Squibb). Coadministration of the angiotensin II receptor antagonist tends to reverse the potassium loss due to the hydrochlorothiazide. It is thought that a decrease in angiotensin II leads to decreased aldosterone secretion which may cause an increase in potassium. The same principle is at work when an ACE inhibitor is used with hydrochlorothiazide as in lisinopril-hydrochlorothiazide (Prinzide®, Merck).

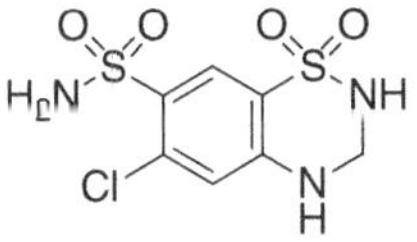

Hydrochlorothiazide

11.4 PROTON PUMP INHIBITORS

Proton pump inhibitors (PPI) are used for people with ulcers and for treatment of heartburn and other symptoms of gastroesophageal reflux disease (GERD), a condition in which backward flow of acid from the stomach causes heartburn and possible injury of the esophagus. GERD has an estimated prevalence of 10–20% in the Western world [32] and proton pump inhibitors are the mainstay of GERD therapy. One study [33] estimated that GERD was prevalent in 7 million insured people in the U.S. at an incremental health cost of $23 billion. Using a 4% work productivity loss, the cost estimate increases to $32 billion.

The stimulation of the proton pump (H^+/K^+-ATPase) in the gastric parietal cell is the final step of acid secretion and proton pump inhibitors block this enzyme [34]. Omeprazole was launched in 1988 as Losec in Europe and then in 1990 as Prilosec in the United States [35]. This drug was followed by lansoprazole, pantoprazole, rabeprazole and esomeprazole, the S enantiomer of omeprazole. PPIs are prodrugs that accumulate in the acidic secretory canaliculi of the gastric parietal calls where they are converted to the sulfenamide derivative which in turn forms a disulfide bond with the thiol group of cysteines in the H^+/K^+-ATPase causing an irreversible inhibition of the proton pump. This is shown with omeprazole [36]. Note that due to the methoxy group on the benzimidazole two isomers of the sulfenamide are formed. Compared with omeprazole, lansoprazole and rabeprazole have different substituents on the pyridine ring and pantoprazole has different substituents on both the pyridine ring and the benzimidazole. The different groups on both the pyridine ring and the benzimidazole influence the pKa of the prodrug and therefore the rate of the conversion to the active metabolite.

Omeprazole is a chiral molecule. There is a chiral center at the sulfur atom of the sulfoxide. Omeprazole exists as a racemic mixture of R and S enantiomers. The active metabolite, the sulfenamide, loses this chiral center and therefore the metabolite is achiral. Researchers studied the enantiomers, originally isolating each by preparing diasteromeric salts with mandelic acid and then separating by chromatography [37]. In humans, the S enantiomer of omeprazole has the highest bioavailability and potency in inhibiting gastric acid secretion due to stereoselective metabolism of omeprazole. In 2000 this finding resulted in the launch of the drug as a single S enantiomer, esomeprazole, known as Nexium®.

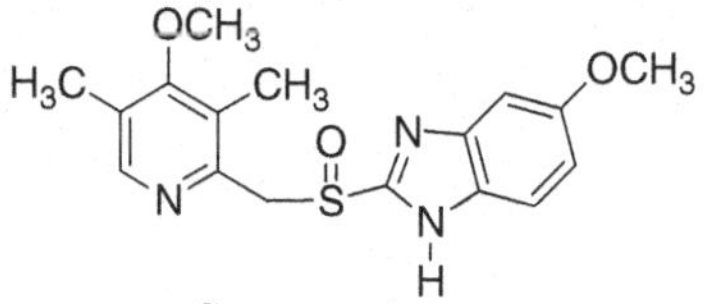

Omeprazole

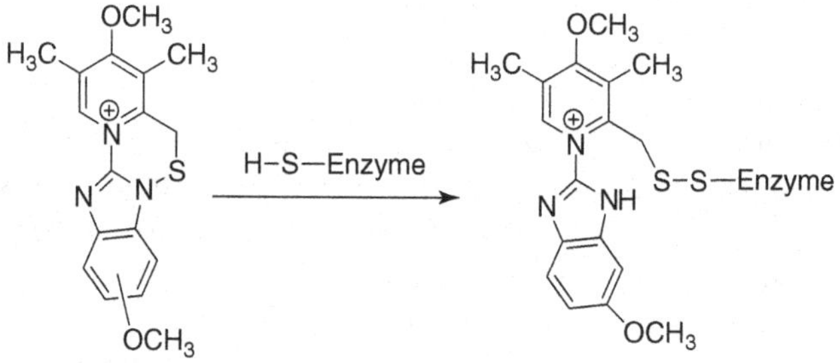

H-S—Enzyme

Sulfenamide derivative of omeprazole

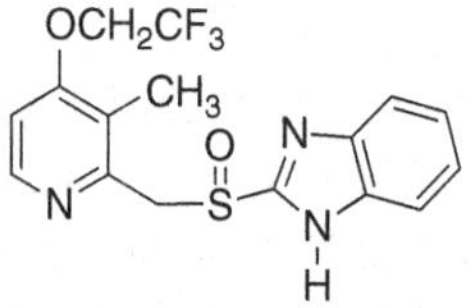

Lansoprazole

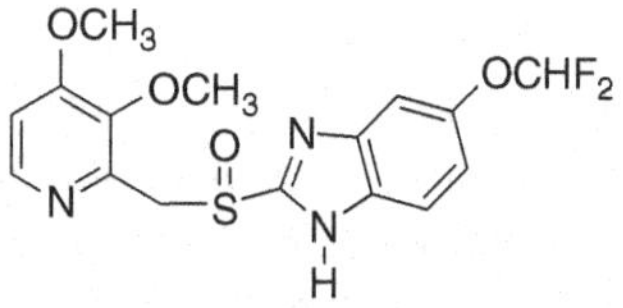

Pantoprazole

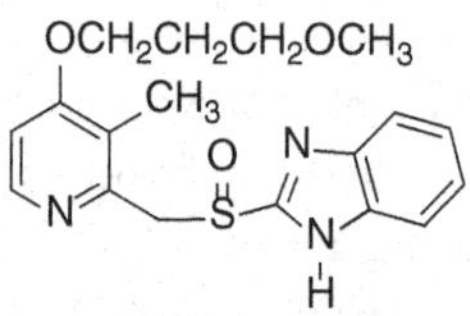

Rabeprazole

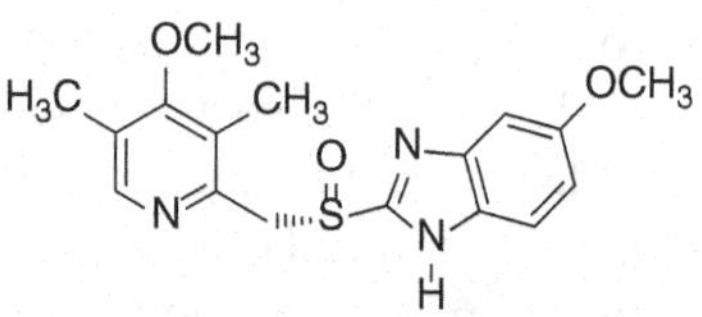

Esomeprazole

Omeprazole can be made by reaction of a substituted diaminobenzene to make the benzoimidazole [38]. Alkylation of the thiol gives the sulfide which is then oxidized to the sulfoxide, omeprazole. The omeprazole can be separated into the two enantiomers. Preferably, asymmetric oxidation of the sulfide is done to selectively prepare the S enantiomer, esomeprazole.

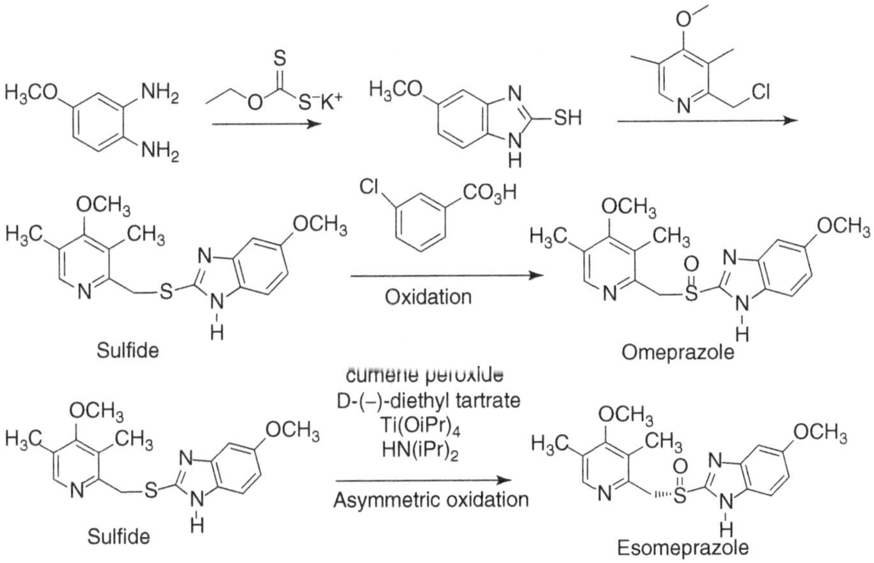

11.5 DIABETES

Diabetes is a disease in which there are high levels of glucose in the blood. It can be caused by too little insulin, a hormone that controls blood sugar, resistance to insulin, or both. Sometimes it is called diabetes mellitus, with mellitus derived from the Latin for honey-sweet. Type 1 involves the body's inability to produce insulin and requires people to inject insulin as a treatment. Gestational diabetes occurs during pregnancy. Among patients diagnosed with diabetes, 90% have Type 2 disease, which is characterized by resistance to insulin action and impairment of insulin secretion, rather than an absolute lack of insulin production, as in patients with Type 1 disease [39]. Estimates are that diabetes affects about 8% of the United States population and 250 million people worldwide; it is the seventh leading cause of death in the United States [40]. There are several drugs used to help control sugar levels and many patients require treatment with more than one type of drug.

Metformin, dimethylbiguanide, also known as glucophage is a common drug prescribed for diabetes. Despite being used for over 50 years and being the most widely prescribed drug for Type 2 diabetes, the mechanism of action remains imperfectly understood [41]. It is linked to an improved peripheral sensitivity to insulin through a stimulated tissue glucose uptake by a transporter-linked system [42]. It also suppresses glucose production. Its discovery can be traced back to medieval times when French lilac was used medicinally. Later, it was discovered that French lilac was rich in guanidine and in the mid 1900s, Jean Sterne explored the antidiabetic properties of several biguanides including clinical studies of dimethylbiguanide, which he termed “glucophage” for “glucose eater [43].”

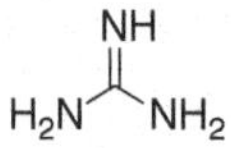

Guanidine 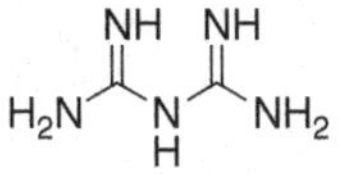 Biguanide 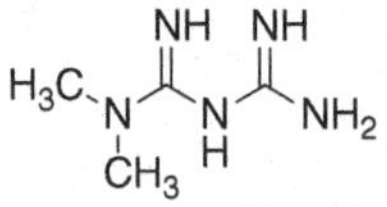 Metformin

Sulfonylureas represent another widely prescribed class of hypoglycemic agents. Sulfonylureas stimulate insulin release from pancreatic β-cells [44]. They bind to a specific receptor, called the sulfonylurea receptor on the β-cell surface [45]. Binding of the sulfonylurea increases calcium influx which in turn triggers insulin release. The structural differences influence the binding capacity for the pancreatic β-cells which can allow for greater potency and lower doses [46]. Perhaps the most common is glimepiride (brand name Amaryl®, Sanofi-Aventis). Tolbutamide, glibenclamide, and glicazide are other examples. Tolbutamide is considered a first-generation sulfonyl urea and others depicted are considered second-generation. Notice that the second-generation sulfonylureas are more lipophilic and this enables them to more easily penetrate the cell membranes.

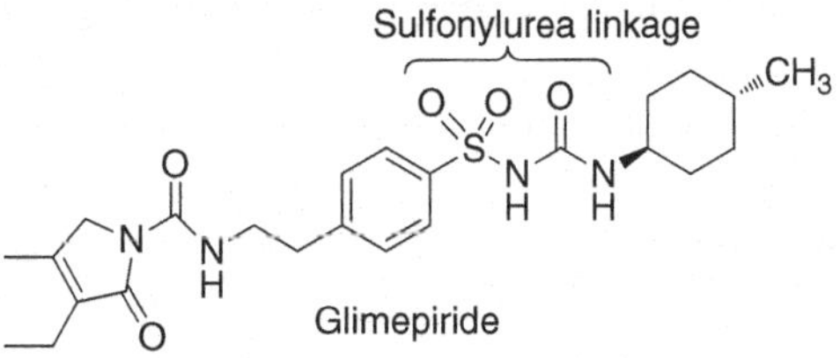

Sulfonylurea linkage

Glimepiride

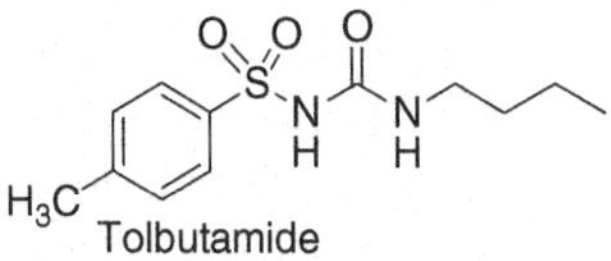

Tolbutamide 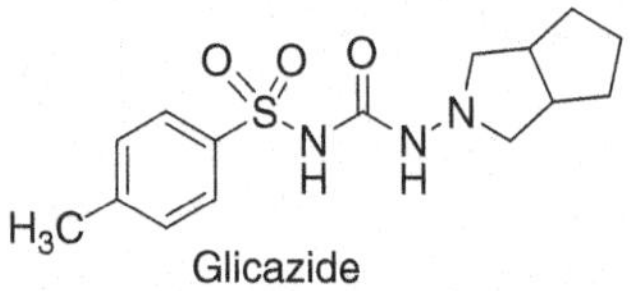 Glicazide

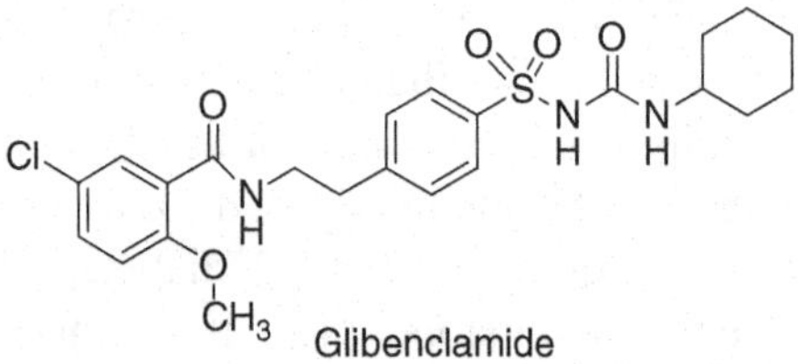

Glibenclamide

Meglitinides are another class of drugs used in the management of Type 2 diabetes. Similar to the sulfonylureas, they bind to the pancreatic β-cells, albeit at a different site, and open the calcium channels. This induces insulin secretion. Repaglinide (Prandin®, Novo Nordisk) is an example of the class.

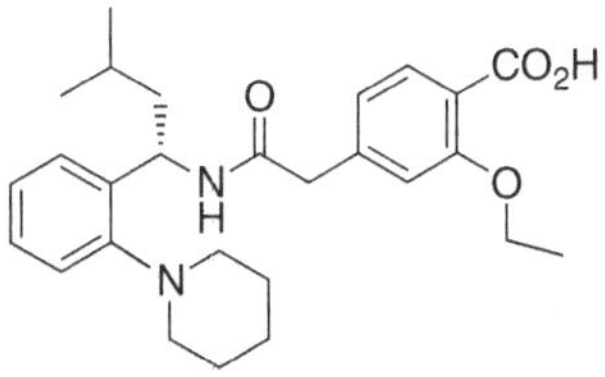

Repaglinide

Thiazolidinediones are a class of insulin-sensitizing drugs. Their use is limited because they can cause congestive heart failure. Troglitazone (Rezulin®) was approved in the U.S. in 1997 followed in 1999 by pioglitazone and rosiglitazone [47]. Troglitazone was withdrawn from the market in 2000 due to liver toxicity. Pioglitazone is available as the hydrochloride salt (Actos®, Takeda) and rosiglitazone as the maleate salt (Avandia®, GlaxoSmithKline). There are also concerns with pioglitazone and rosiglitazone (risk of cardiac events) and this precludes their use in some patients.

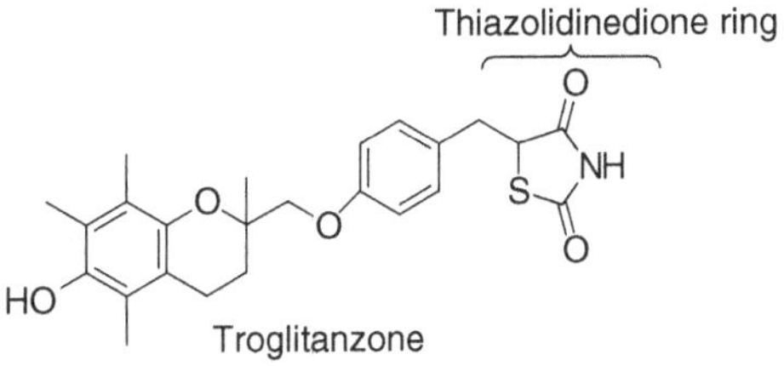

Thiazolidinedione ring

Troglitanzone

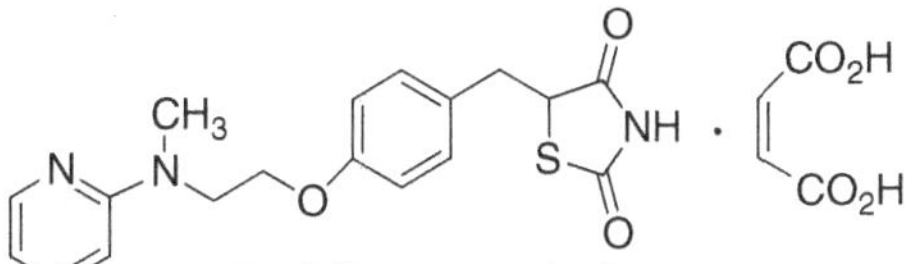

Rosiglitazone maleate

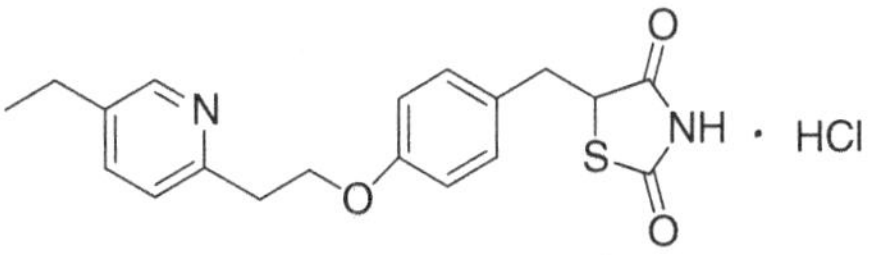

Pioglitazone hydrochloride

Notice that there is a chiral carbon in these structures. They are present as a racemic mixture of enantiomers. The enantiomers interconvert rapidly in vivo and there is no difference in the pharmacologic activity between the two enantiomers.

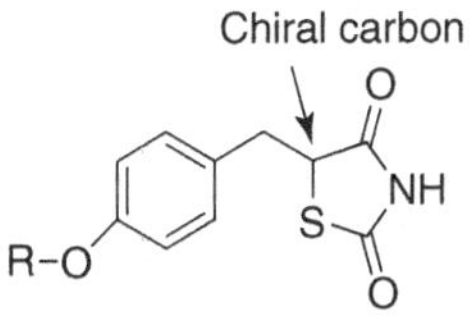

Chiral carbon

Incretin hormones such as glucagon-like peptide 1 (GLP-1) cause insulin release from the pancreatic β-cells. Exenatide (Byetta®, Amylin

Pharmaceuticals) was approved in the U.S. in 2005 and is an injectable 39-amino acid synthetic peptide. It was originally identified in the saliva of the Gila monster (*Heloderma suspectum*). Exenatide mimics the blood-sugar lowering activity of glucagon-like peptide 1 (GLP-1) [48]. It affects pancreatic beta-cell responsiveness to glucose. This leads to insulin release, predominantly in the presence of elevated glucose concentrations. Liraglutide (Victoza®, Novo Nordisk) is an analog of GLP-1. It was approved in the U.S. in 2010. It is also a peptide and, like exenatide, promotes insulin release in the presence of elevated glucose levels.

The DPP-4 enzyme inactivates incretin hormones such as GLP-1. Some diabetes drugs work by inhibiting DPP-4 and therefore slow the inactivation of incretin hormones. Sitagliptan (the phosphate salt is marketed as Januvia®, Merck) and saxagliptin (Onglyza®, Bristol-Myers Squibb) are examples of DPP-4 inhibitors.

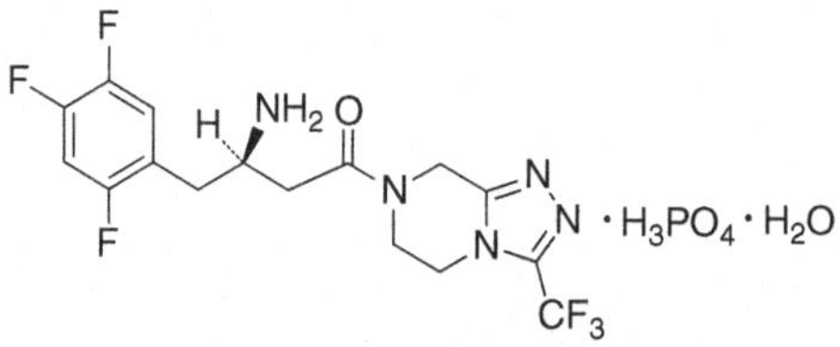

Sitagliptan phosphate monohydrate

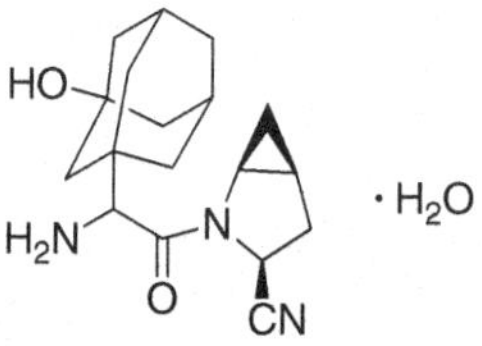

Saxagliptan monohydrate

11.6 ANTIDEPRESSANTS

Major depressive disorder is among the most common psychiatric disorders in the U.S. with an estimated prevalence of about 10% in the general population [49]. The effects of antidepressants can be linked to the improvement of neurotransmission by the monoamine neurotransmitters, serotonin, norepinephrine, and dopamine [50].

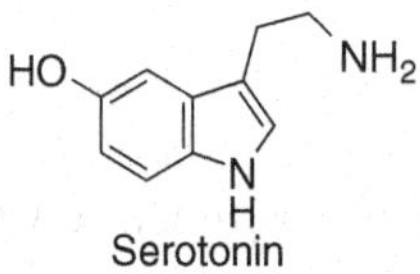

Serotonin

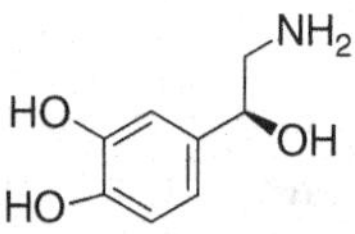

Norepinephrine

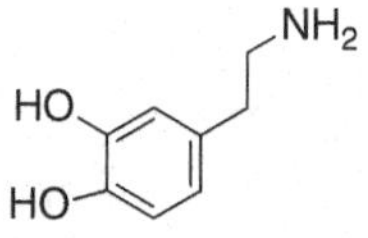

Dopamine

Early antidepressant medications were called tricyclic antidepressants (TCA), exemplified by dothiepin and amitriptyline. However, there are toxicity issues with these medications. The next generation of antidepressants was the monoamine oxidase inhibitors (MAOIs). As their name implies, they inhibit the monoamine oxidase enzyme thereby inhibiting the oxidation of monoamines such as serotonin that act as neurotransmitters. Examples include selegiline (available in transdermal patch form as Emsam®, Somerset

Pharmaceuticals, Inc.) and isocarbazid (Marplan®, Validus Pharmaceuticals). They also suppress tyramine uptake. In response to diet-related surges of tyramine, some patients experienced sudden increases in blood pressure that caused fatal brain hemorrhages [51]. Newer antidepressant medications are based upon the inhibition of serotonin reuptake.

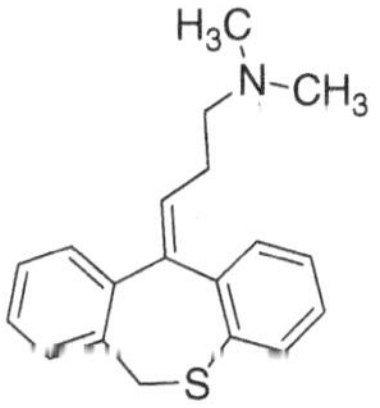

Dothiepin 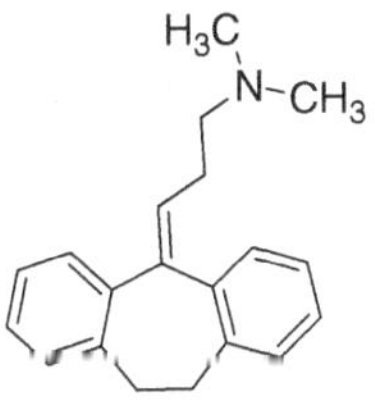 Amitriptyline

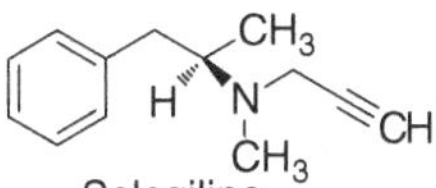

Selegiline 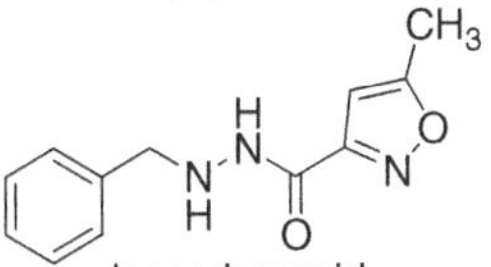 Isocarboxazid

Serotonin, also known as 5-hydroxytryptamine (5-HT) is biosynthesized from tryptophan and is a neurotransmitter. Serotonin plays an important role in many behaviors including sleep, appetite, memory, and mood [52]. People with depressive disorders exhibit low levels of serotonin in the synapses. Protonated serotonin binds to a serotonin reuptake transporter protein, sometimes referred to as the serotonin transporter (SERT) and is then moved to an inward position on the neuron and subsequently released into the cytoplasm. Selective serotonin reuptake inhibitors (SSRI) bind with high affinity to the serotonin binding site of the transporter. This leads to antidepressant effects by increasing extracellular serotonin levels which in turn enhances serotonin neurotransmission [53]. The SSRI class of antidepressants has fewer side effects than the monoamine oxidase inhibitors.

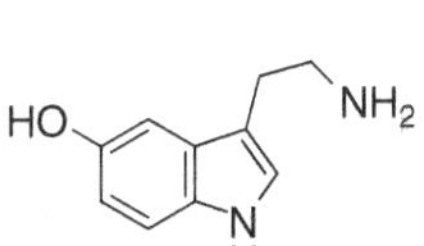

Serotonin 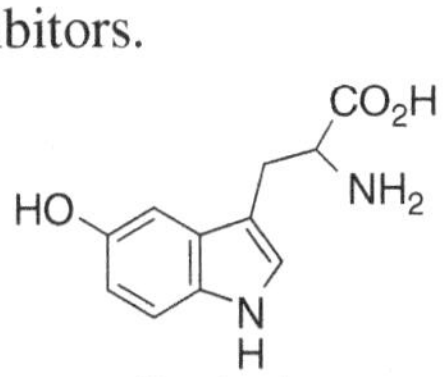 Tryptophan

There are several SSRI inhibitors. Their chemical structures are different but they have similar modes of action, namely binding to the SERT. However, they have different pharmacokinetic parameters such as half-life, and differences in adverse effects and drug interactions [54]. Fluoxetine was the first drug of the class, approved in the U.S. in 1987. It is a chiral molecule and the racemic mixture is used as the hydrochloride salt (Prozac®, Lilly). It has been approved and marketed in more than 90 countries and used by more than

54 million people worldwide [55]. Scientists at Eli Lilly and company recognized that antihistamines had potential as antidepressants. They began synthesizing analogs based upon the antihistamine diphenyhydramine (Benadryl®, Johnson and Johnson) and found fluoxetine to be a selective serotonin reuptake inhibitor [56]. Eli Lilly scientists believed that the enantiomers were not significantly different, but in the mid-1990s, the drug company Sepracor obtained separate patents for use of (*S*)- and (*R*)-fluoxetine to treat migraine and depression, respectively. In 1998, Eli Lilly and Sepracor agreed to codevelop (*R*)-fluoxetine as the single-isomer, side-effect-free version of Prozac, but the agreement was terminated in 2000 when clinical studies showed adverse side effectsat high doses not observed with the raccmate [57].

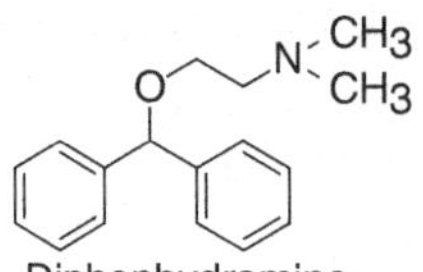

Diphenhydramine 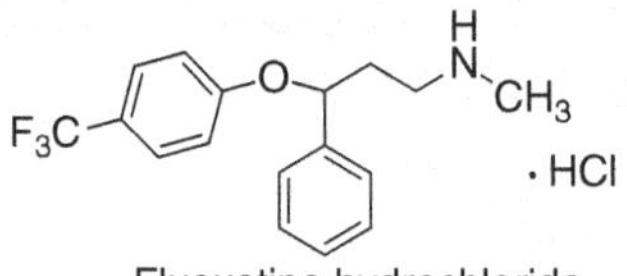   Fluoxetine hydrochloride

Other SSRIs include paroxetine hydrochloride (Paxil®, Glaxo SmithKline), fluoxamine maleate (Luvox®, Abbott), sertraline hydrochloride (Zoloft®, Pfizer), vilazodone hydrochloride (Viibryd®, Forest Laboratories), and citalopram hydrobromide (Celexa®, Forest Laboratories).

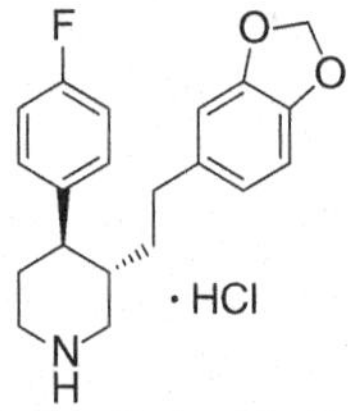

Paroxetine hydrochloride 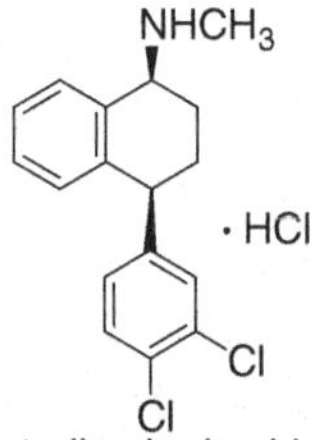   Sertraline hydrochloride 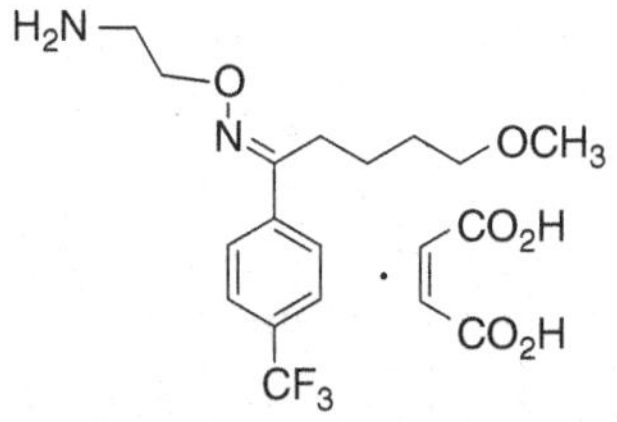   Fluoxamine maleate

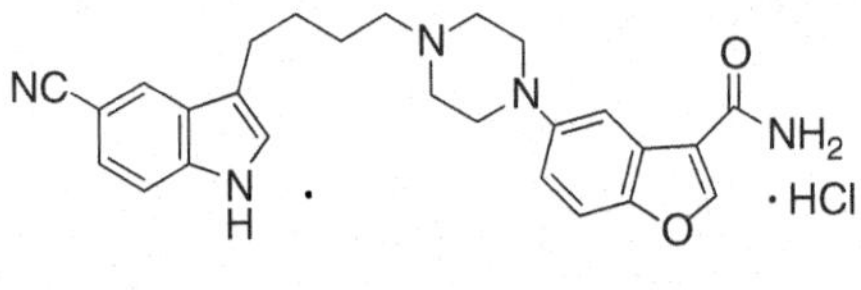

Vilazodone hydrochloride 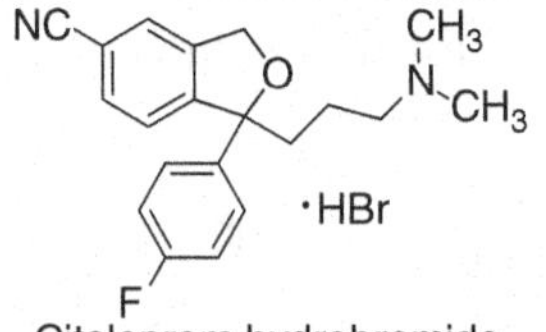   Citalopram hydrobromide

Paroxetine has two chiral carbons and is administered as a single stereoisomer. Citalopram is one chiral carbon and is administered as a racemic mixture in Celexa® and as the pure S enantiomer in escitalopram oxalate (Lexapro®, Forest Laboratories).

In 2004, 17 years after the introduction of Prozac®, duloxetine hydrochloride (Cymbalta®, Lilly) was approved in the U.S. Duloxetine has a chiral carbon and is the S enantiomer. Duloxetine is a dual reuptake inhibitor of serotonin and norepinephrine (SSNRI). It is effective against depression and is also prescribed for persistent pain. Venlafaxine hydrochloride (Effexor®, Pfizer) is also a dual reuptake inhibitor of serotonin and norepinephrine. Venlafaxine has a chiral carbon and is administered as a racemic mixture.

Duloxetine hydrochloride Venlafaxine hydrochloride

REFERENCES

1. Louis Quin, John Tyrell. *Fundamentals of Heterocyclic Chemistry*. New York: Wiley; 2010:196.
2. D Johnson, J Li, editors. *The Art of Drug Synthesis*. New York: Wiley; 2007.
3. J Li, D Johnson, D Sliskovic, B Roth. *Contemporary Drug Synthesis*. New York: Wiley; 2004.
4. J Saunders. *Top Drugs*, Top Synthetic Routes. Oxford University Press; 2000.
5. T Lemke, et al., editors. *Foye's Principles of Medicinal Chemistry*. 7th ed. Lippincott, Baltimore, MD; 2013:1031.
6. Abbott 2011 Annual Report, Abbott Laboratories, Abbott Park, IL.
7. N Singh, et al. Mini-Reviews in Medicinal Chemistry 2009; 9:1272–1283.
8. V Roger, et al. Circulation 2012; 125:e2–e220.
9. E Istvan, J Deisenhofer. Biochimica et Biophysica Acta 2000; 1529:9–18.
10. E Istvan. Atherosclerosis Supplements 2003; 4:3–8.
11. Irina Buhaescu, Hassane Izzedine. Clinical Biochemistry 2007; 40:575–584.
12. N. Singh, et al. Mini-Reviews in Medicinal Chemistry 2009; 9:1272–1283.
13. John LaMattina. *Drug Truths*. New York: Wiley; 2009:15.
14. AstraZeneca Annual Report and Form 20-F Information 2011.
15. R Monaghan, et al. U.S. Pat. No. 4,231,938. 1980.
16. W Hoffman, R Smith, A Willard. U.S. Pat. No. 4,444,784. 1984.
17. VA O'Neill, KL Evenstad. U.S. Pat. No. 5,268,181. 1993.
18. D Johnson, J Li, editors. *The Art of Drug Synthesis*. New York: Wiley; 2007:187.
19. Sherry L Murphy, Jiaquan Xu, Kenneth D Kochanek. National Vital Statistics Reports January 11, 2012; 60(4).
20. V Roger, et al. Circulation 2012; 125:e2–e220.
21. V Roger, et al. Circulation 2012; 125:e2–e220.
22. T Lemke, et al., editors. *Foye's Principles of Medicinal Chemistry*. 7th ed. Lippincott, Baltimore, MD; 2013:784.
23. D Carini, J Duncia, P Wong. U.S. Pat. No. 5,138,069. 1992.
24. D Johnson, J Li, editors. *The Art of Drug Synthesis*. New York: Wiley; 2007:132.

25. W Cabri, R DiFabio. *From Bench to Market.* Oxford University Press; 2000:120–145.
26. D Johnson, J Li, editors. *The Art of Drug Synthesis.* New York: Wiley; 2007:130.
27. C Smith, J Vane. The FASEB Journal 2003; 17(8):788–789.
28. M Remko. Chem. Pap. 2007; 61(2):133–147.
29. D Johnson, J Li, editors. *The Art of Drug Synthesis.* New York: Wiley; 2007:148.
30. T Lemke, et al., editors. *Foye's Principles of Medicinal Chemistry.* 7th ed. Lippincott, Baltimore, MD; 2013:772.
31. J Saunders. *Top Drugs, Top Synthetic Routes.* Oxford University Press; 2000:22.
32. E Altan, et al. Expert Opin. Emerging Drugs 2012; 17(3):347.
33. RA Brook, et al. Ailment Pharmacology and Therapeutics 2007; 26:889.
34. Silvia Marelli, Fabio Pace. Expert Rev. Gastroenterol. Hepatol. 2012; 6(4):423.
35. L Olbe, E Carlsson, P Lindberg. Nature Reviews Drug Discovery 2003; 2:132–139.
36. A Brandstrom, et al. Acta Chemica Scandinavica 1989; 43:536.
37. L Olbe, E Carlsson, P Lindberg. Nature Reviews Drug Discovery 2003; 2:132–139.
38. J Li, D Johnson, D Sliskovic, B Roth. *Contemporary Drug Synthesis.* New York: Wiley; 2004:23–25.
39. Stephen N Davis. Journal of Diabetes and Its Complications 2004; 18:367– 376.
40. Johanna DiStefano, Richard Watanabe. Pharmaceuticals 2010; 3:2610–2646.
41. Russell A Miller, Qingwei Chu, Jianxin Xie, Marc Foretz, Benoit Viollet, Morris J. Birnbaum. Nature 2013; 494:256–260.
42. C Sirtori, C Pasik. Pharmacol Res. 1994; 30(3):187–228.
43. C Bailey, C Day. Pract. Diab. Int. 2004; 21(3):115–117.
44. Johanna DiStefano, Richard Watanabe. Pharmaceuticals 2010; 3:2610–2646.
45. Mary Korytkowski Pharmacotherapy 2004, 24(5):606–620.
46. Stephen N Davis. Journal of Diabetes and Its Complications 2004; 18:367– 376.
47. Johanna DiStefano, Richard Watanabe. Pharmaceuticals 2010; 3:2610–2646.
48. Lisa Jarvis. Chemical and Engineering News 2012, 90(28):7.
49. J Li, D Johnson, D Sliskovic, B Roth. *Contemporary Drug Synthesis.* New York: Wiley; 2004:125.
50. D Johnson, J Li, editors. *The Art of Drug Synthesis.* New York: Wiley; 2007:200.
51. D Johnson, J Li, editors. *The Art of Drug Synthesis.* New York: Wiley; 2007:201.
52. Lino Sghendo, Janet Mifsud J Pharmacy and Pharmacology 2011; 64:317–325.
53. H Zhong, N Haddjeri, C Sanchez. Psychopharmacology 2012; 219:1–13.
54. Lino Sghendo, Janet Mifsud J Pharmacy and Pharmacology 2011; 64:317–325.
55. Melissa Braddock. Chemical and Engineering News 2005; 90(25):104.
56. B Molloy, K Schmiegel. U.S. Pat. No. 4,018,895. 1977.
57. AM Rouhi. Chemical and Engineering News 2003; 81(18):56–61.

CHAPTER 12

Agricultural Chemicals

12.1 OVERVIEW

Global population growth is about 1.3% every year, and the population is expected to reach seven billion by 2015 and nine billion by 2050 [1]. In the time it takes to read this sentence, another 20 people will have been added to the world's population [2]. The number of malnourished people in the world has been estimated at 852 million. Each year, hunger and malnutrition are responsible for the deaths of six million children. In want of food, people plow forests and the world annually loses 9.4 million hectares of forests [3]. A hectare (ha) is 10,000 square meters and 9.4 million ha is about 36,000 square miles or about the size of the entire state of Indiana. Exacerbating the food problem is the use of crops as fuel. For example, in the United States, corn is used to make ethanol for gasoline. In Brazil, sugar cane is used. There are other efforts to use crops such as soybean for biodiesel. It has been estimated that food production will have to increase 70% by 2050 to feed the growing population [4].

"Pesticides" is a general term and includes substances that kill weeds (herbicides), insects (insecticides) and fungus (fungicides). Although the terms might be "bad words" in some circles, it is only because of the use of these agricultural chemicals that the problem of food supply and deforestation is not worse. Agricultural chemicals, along with improved farming techniques, give us hope that we will be able to feed the increased population in the future.

It takes about 11 years to bring a new agricultural chemical from discovery to the market and one estimate has the cost at about $200 million [5]. Another estimate is at an R&D price tag of $265 million, with only one of 140,000 potential candidates making it to market [6]. The global agricultural chemicals market is about $38 billion [7].

Fundamentals of Industrial Chemistry: Pharmaceuticals, Polymers, and Business, First Edition. John A. Tyrell.

Six multinational companies: Syngenta, Bayer, Monsanto, DuPont, BASF, and Dow are major players in the agrochemical industry [8]. Each has annual crop protection chemical sales in excess of $3 billion and together they have about 70% of the market [9]. China is emerging as first in world pesticide production and second in usage with about 2,800 manufacturers and more than 400 million small-scale farmers [10]. There is great variety among pesticides. For example, in Canada there are over 7,000 pesticide products and over 500 active ingredients are registered [11]. More than 80% of the pesticides used in Canada are herbicides [12].

12.2 FERTILIZER

Fertilizers are compounds given to plants to promote growth [13]. Plants require elemental nutrients in various amounts, which vary with species, genotype, soil, and environmental factors [14]. For grain crops, as seed is produced and removed from the field, nutrients are lost every season and must be replaced. Fertilizers are used for this purpose. Nitrogen, phosphorus, and potassium are primary nutrients. Primary nutrients are required in large amounts and are quickly depleted from the soil. Retail fertilizers for homeowners typically have a three-number designation such as 10-6-4. These numbers designate the weight percent of nitrogen, phosphorus, and potassium. The first number is the weight percent nitrogen in the fertilizer, 10% in this example. The second number is for phosphorus and represents the amount on a phosphate basis. In this example, there is 6% P_2O_5. Recognize that the phosphorus is not present as phosphorus pentoxide; this is a way of reporting the number. The third number is for potassium on a potash or potassium oxide basis. Fertilizer having the designation of 10-6-4 has the amount of potassium equivalent to 4% K_2O. In addition to these three elements, many other elements are also needed for proper plant growth albeit often in smaller quantities.

Nitrogen is usually the most limiting nutrient in crop protection, partly because large amounts are required by plants and partly because it migrates from the soil [15]. Common nitrogen fertilizers are ammonia and derivatives of ammonia such as urea, ammonium nitrate, and ammonium phosphate.

To understand the role of fertilizers, it might be helpful to think about soybean. Soybean is the world's most important source of protein and accounts for nearly 70% of the protein meal consumption and nearly 30% of vegetable oil consumption [16]. Nitrogen, phosphorus, and potassium (in that order) are removed in soybean seed and need to be replenished. Soybean is a source of protein and protein is about 16% nitrogen [17]. Phosphorus is used in key constituents such as adenosine triphosphate (ATP), which is used in photosynthesis. Potassium serves as an important cation for various plant processes and is required for plant growth. Nitrogen, phosphorus, and potassium are sometimes called macronutrients. Other elements required include sulfur, calcium, and magnesium; sometimes these are called secondary macronutrients. Many other elements are also required but in smaller quantities and these are called micronutrients [18].

12.3 INSECTICIDES

The Insecticide Resistance Action Committee (IRAC) was formed in 1984 to promote the development of insecticide resistance management strategies with the ultimate goal of sustainable agriculture and improved public health [19]. IRAC has classified insecticides based upon their mode of action (MoA) in 30 groups (28 groups plus a group for unknown action and another group for non-specific action).

Organophosphates and carbamates are in group 1, acetylcholinesterase inhibitors. Organophosphate insecticides are potent and effective anti-cholinergic agents and represent the largest class of insecticides sold worldwide [20]. Most organophosphates are not persistent and do not bioaccumulate [21]. Through their inhibition of acetylcholinesterase, the enzyme responsible for catalyzing the breakdown of the neurotransmitter acetylcholine, the organophosphates are highly toxic to insects. This chemical class includes over 200 pesticides and includes diazinon, chlorpyrifos, and malathion [22]. Malathion is the most common organophosphate insecticide applied in the United States [23]. Diazinon and chlopyrifos are not approved for residential use in the United States. Chlorpyrifos demand is growing in Asia, especially in China where it is replacing more toxic organophosphates [24].

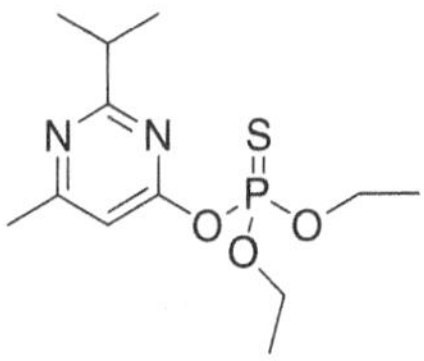

Diazinon 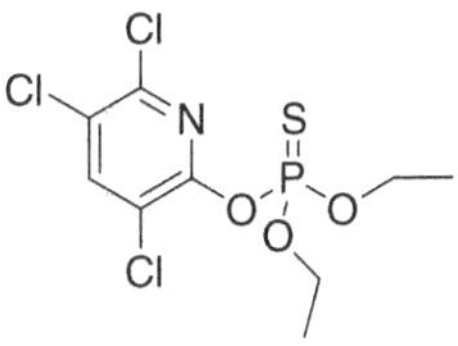 Chlorpyrifos 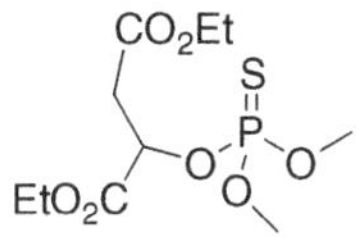 Malathion

Carbamates include aldicarb, methiocarb, methomyl, carbofuran, bendiocarb and oxamyl. Although they are broad-spectrum insecticides, with moderate toxicity and persistence, they rarely bioaccumulate or causc major environmental impacts [25].

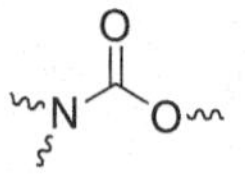

Carbamate linkage

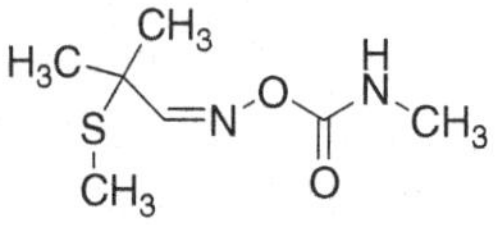

Aldicarb

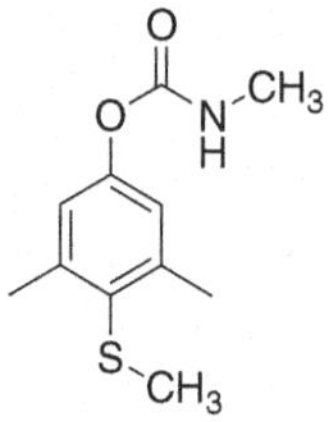

Methiocarb

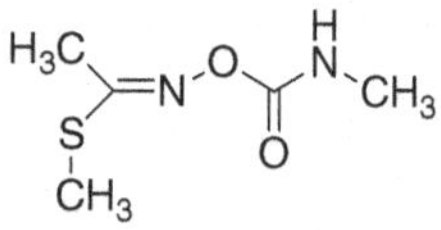

Methomyl

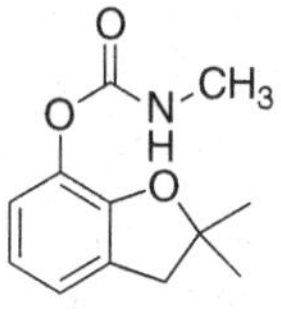

Carbofuran

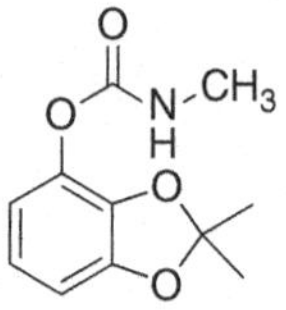

Bendiocarb

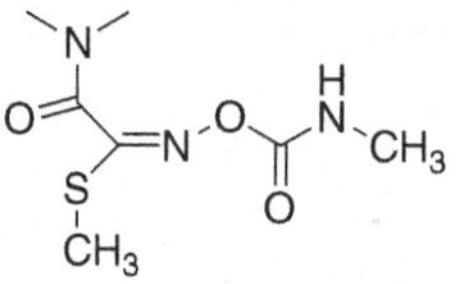

Oxamyl

Pyrethroid insecticides are widely used because of their high activity as an insecticide and low mammalian toxicity. Pyrethroids are in group 3, sodium channel modulators. The pyrethroids have a highly nonpolar nature, low water solubility, and high affinity to soil and sediment particulate matter. Natural pyrethrin is extracted from the flowers of *Chrysanthemum* spp., and its use was already known in China in the first century A.D. Pyrethroids, synthetic analogues of pyrethrin, have been produced since 1940 [26].

The six pyrethrins, Pyrethrin I, Jasmolin I, Cinerin I, Pyrethrin II, Jasmolin II, and Cinerin II were found to be present in the leaf extract of the *Chrysanthemum cinerariaefolium* seedlings. Pyrethrins are thought to act as defense mechanisms that prevent insects from feeding on the plant [27]. The pyrethrins repel insects and also paralyze flying insects, thereby exhibiting a "knock-down" effect. The active components occur to a small extent in all parts of the plant, but especially in the flower heads [28].

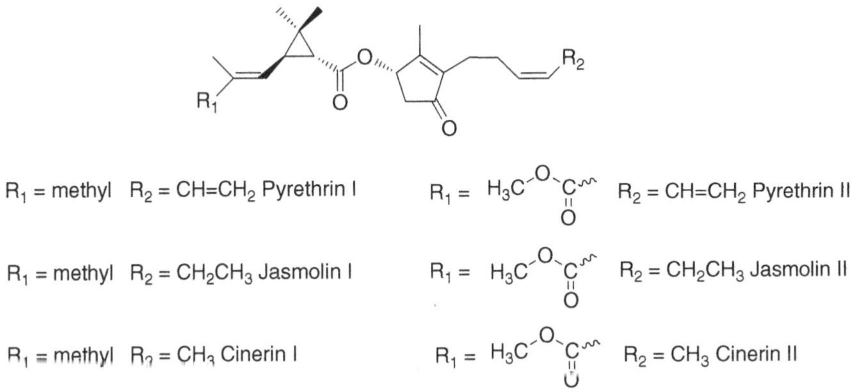

R_1 = methyl R_2 = CH=CH_2 Pyrethrin I | R_1 = H_3C–O–C(=O)– R_2 = CH=CH_2 Pyrethrin II

R_1 = methyl R_2 = CH_2CH_3 Jasmolin I | R_1 = H_3C–O–C(=O)– R_2 = CH_2CH_3 Jasmolin II

R_1 = methyl R_2 = CH_3 Cinerin I | R_1 = H_3C–O–C(=O)– R_2 = CH_3 Cinerin II

Pyrethroids often have two, three or more chiral centers and therefore can have four, eight, or more stereoisomers. Sometimes they are sold as single isomers, such as deltamethrin, useful for more than 150 crops including vegetables, fruits, cereals, oil seed rape, rice, soybeans, and corn. This pyrethroid has eight possible stereoisomers. The chiral carbons are designated by an asterisk in the depicted structure. The commercial product has R configuration at each of the chiral cyclopropyl carbons and S at the cyanobenzylic carbon. Deltamethrin is highly toxic to insects. Compared with pyrethrin I, it is 1400 times as toxic to the housefly [29].

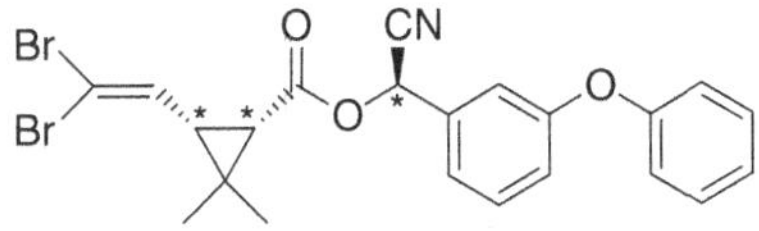

Deltamethrin

Cypermethrin has eight possible stereoisomers; it is used as a mixture of stereoisomers but only two of the eight are active. Typically the R configuration at the cyclopropyl carbon alpha to the carbonyl of the ester is more active than the S configuration; for example in permethrin it is 25 times more toxic to houseflies than the R configuration [30].

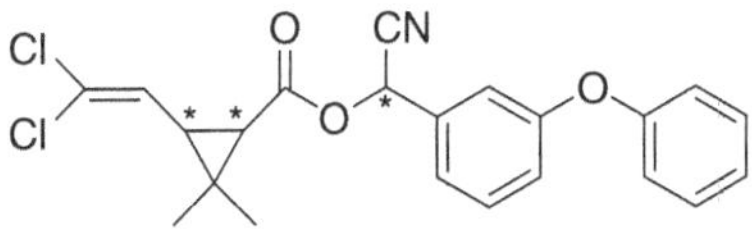

R more toxic to houseflies

Cypermethrin

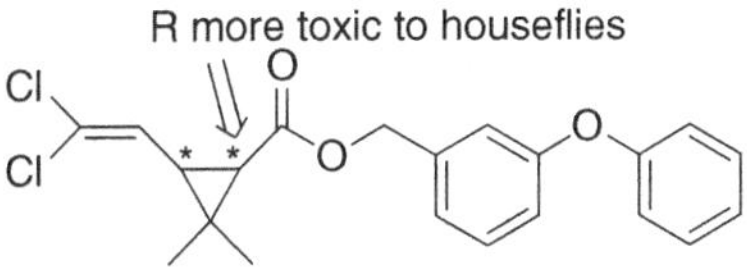

Permethrin

Most mosquito devices such as coils, mats, or vaporizers contain pyrethroids because mosquitoes are rapidly paralyzed and cannot suck blood, and because of the low mammalian toxicity of the pyrethroids. The pyrethroids are characterized as having a cyclopropane carboxylic acid ester functionality. Some commercial pyrethroids that are used in mosquito-control devices include d-allethrin, furamethrin, prallethrin, transfluthrin, and metofluthrin [31].

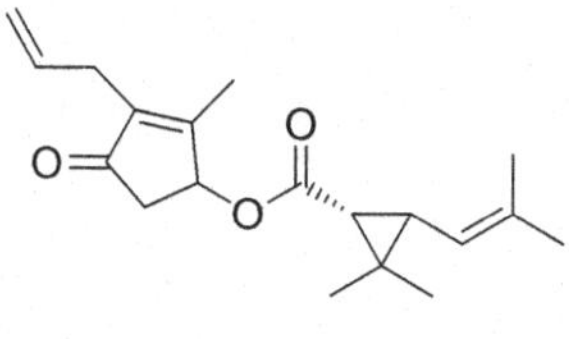

d-allethrin

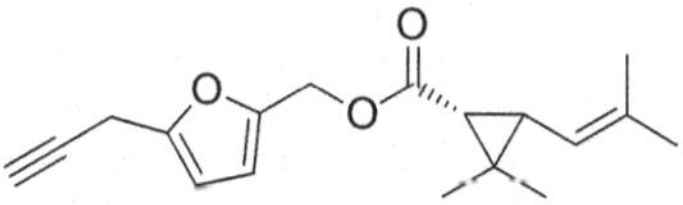

Furamethrin

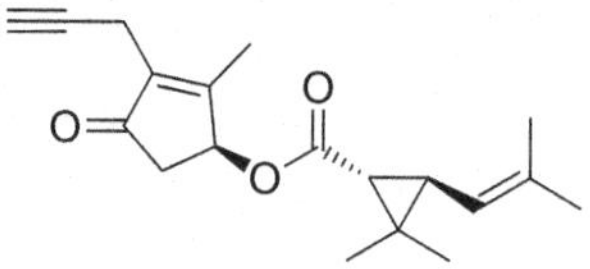

Prallethrin

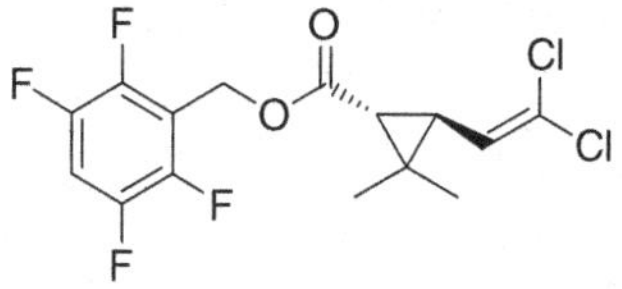

Transfluthrin

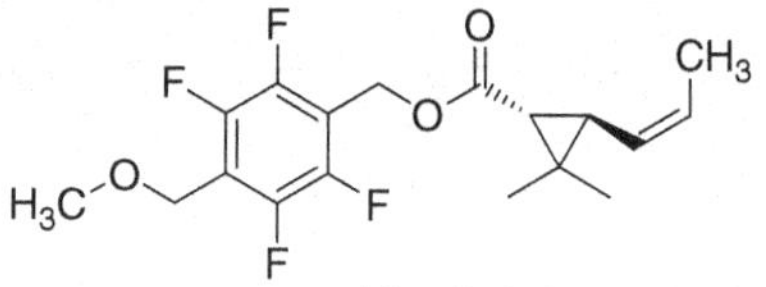

Metofluthrin

By synthesizing hundreds of structural analogs and with several structure activity relationship (SAR) studies, there has been an expansion in the number of synthetic pyrethroids on the market. Type I pyrethroids, represented by, for example, permethrin cause hyperactivity and incoordination in the insect [32]. Type II pyrethroids do not have the cyclopropane ring and contain an alpha-cyanobenzyl ester. Fenvalerate is an example. The Type II pyrethroids induce paralysis. Note that fenvalerate has two asymmetric carbons at the two benzyl positions so there are four possible stereoisomers. It is sold as a mixture of the four stereoisomers and also as the most active, SS stereoisomer, which is called esfenvalerate.

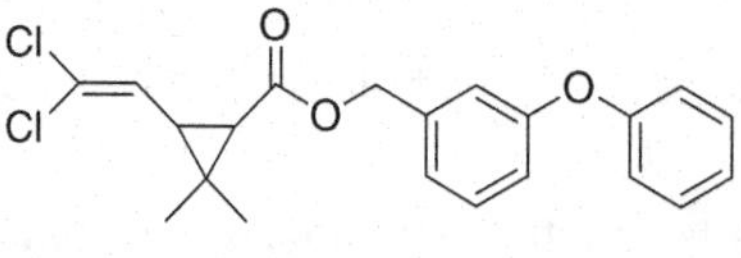

Permethrin

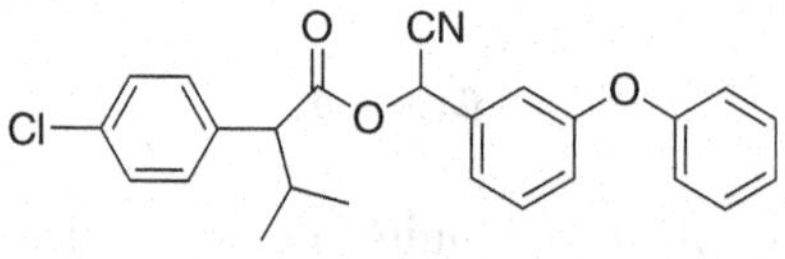

Fenvalerate

Another biological target for many insecticides is the nervous system of the insect because this can result in selective toxicity. The neonicotinoids, in group 4, work by this mechanism and act on insect nicotinic acetylcholine receptors (nAChR). Nicotine, the namesake of the nAChR has been used for pest control since the 17th century [33]. However nicotine is hazardous to people and has limited effectiveness as an insecticide. Nicotine and its analog epibatidine, isolated from a tropical poisonous frog, exist mainly in the protonated form at physiological pH. They are examples of the nicotinoid family and are toxic to mammals.

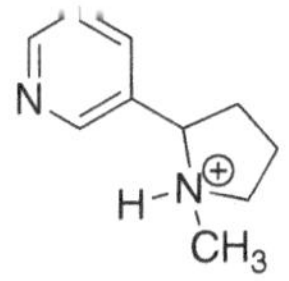

Protonated nicotine

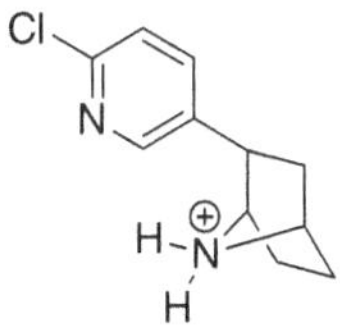

Protonated epibatidine

In contrast, the neonicotinoids are not protonated and therefore bind differently, resulting in greater activity on insects than mammals. It is this selective toxicity that has led to extensive use of the neonicotinoids and they represent about 20% of the global insecticide market [34]. Worldwide sales of neonicotinoid insecticides are estimated at $1 billion [35]. Imidacloprid and thiacloprid are examples of neonicotinoid insecticides.

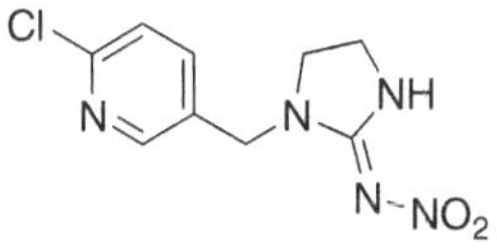

Imidacloprid

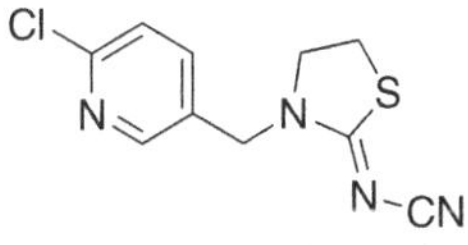

Thiacloprid

Another neonicotinoid example is clothianidin, which is used in the treatment of seeds. Clothianidin is used for canola, cereals, sunflowers, sugar beet, and corn, including 90% of the United States corn crop [36].

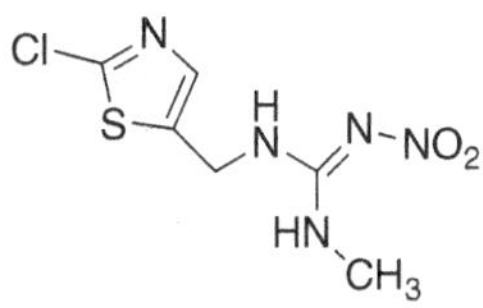

Clothianidin

There is some concern that neonicotinoids may interfere with the growth and viability of beehives [37]. Because bees pollinate, thereby playing a key role in crop growth, this is a concern for agriculture and it is under investigation. Although the bee population has been in decline, there is no clear evidence implicating the neonicotinoids. There are several other possible factors including a parasitic mite that may be the cause. Nosema bombi is a parasite commonly found in bees. One study showed that declining bee populations are associated with both high Nosema bombi infection levels and low genetic diversity [38]. Nonetheless, governments have taken steps to protect bees from possible harm. The European Commission has announced a two-year ban on clothianidin, imacloprid, and thiamethoxam [39]. The EPA is requiring manufacturers of imidacloprid, dinotefuran, clothianidin, and thiamethoxam to include language prohibiting use when bees are foraging or when plants are flowering [40].

There are organochlorine insecticides such as DDT, a group 3, sodium channel modulator. Others such as chlordane, dieldrin, and endosulfan are members of group 3, GABA-gated chloride channel antagonists. Because of their persistence in, and detrimental effect on, the environment, especially on birds, they are no longer widely used in most countries.

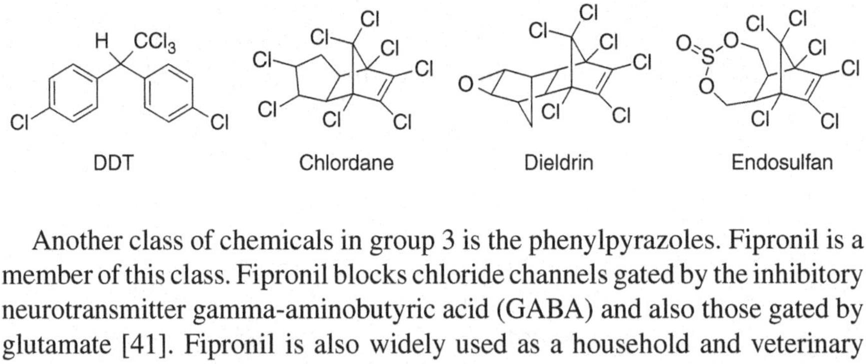

DDT Chlordane Dieldrin Endosulfan

Another class of chemicals in group 3 is the phenylpyrazoles. Fipronil is a member of this class. Fipronil blocks chloride channels gated by the inhibitory neurotransmitter gamma-aminobutyric acid (GABA) and also those gated by glutamate [41]. Fipronil is also widely used as a household and veterinary insecticide for the control of insects such as termites, cockroaches, and ants. As with some neonicitinoid pesticides, there is a concern with the effect of fipronil on honeybees. The European Union has voted in favor of restricting the use of fipronil on seeds to only seeds sown in greenhouses [42].

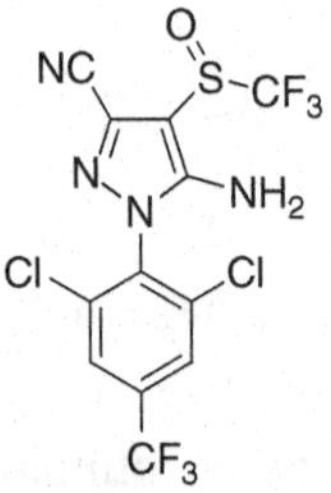

Fipronil

Although it is not typically considered as an agricultural chemical, N,N-diethyl-3-methylbenzamide (DEET) remains as an important insect repellent, especially for mosquitoes. About 200 million people use it every year and over 8 billion doses have been applied over the past 50 years [43]. Estimates are that DEET is used by approximately 30% of the United States population [44].

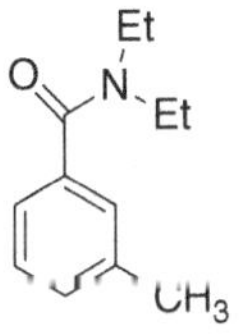

N,N-diethyl-m-toluamide
DEET

Calcium channels represent an attractive biological target for insect control because of their role in multiple cell functions such as muscle contraction. One new insecticide [45] that causes release of intracellular Ca^{2+} is Rynaxypyr™. It is in group 28, ryanodine receptor modulators. The ryanodine receptors contract the muscles; the contraction requires a release of calcium ions. Depletion of the calcium ion results in insect death.

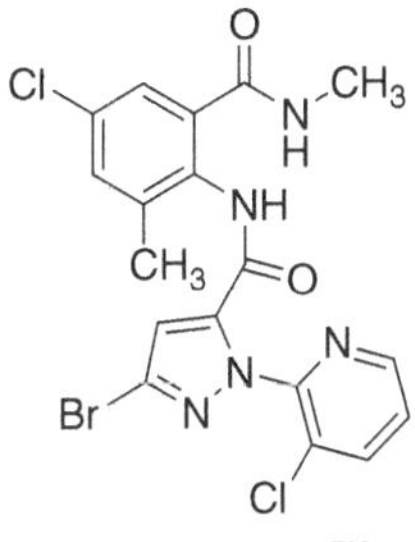

Rynaxypyr™

12.4 HERBICIDES

On a global basis, about 250 species are sufficiently troublesome to be termed weeds, representing approximately 0.1% of the world's flora [46]. Weeds reduce crop yield and therefore result in a need to plant larger acreage for the same crop production. Weeds can also reduce crop quality. Herbicides are "weed-killers' and are used to control unwanted vegetation that hinders the growth of the desired crop.

Weeds can develop resistance to herbicides. Because of this, herbicides are often rotated from season to season. For example, target site resistance can develop where a mutation occurs in the weed that prevents the herbicide from binding. Herbicides operate by different modes of biological action. In order to facilitate rotation of herbicides as a means of managing herbicide resistance, they have been listed based upon their mode of action. When users switch herbicides, they can switch to one with a different mode of action. The Weed Science Society of America (WSSA) has divided them into groups and assigned numbers from 1–27 for each group [47]. For example, group 1 herbicides inhibit acetyl CoA carboxylase. The international Herbicide Resistance Action Committee (HRAC) has a similar classification system using letters and letter–number combinations. In the HRAC system, group 1 herbicides are labeled as group A. They are not all in the same order. Group 3, inhibitors of microtubule assemble, is group K1 in the HRAC system.

The shikimate pathway involves the biosynthesis of aromatic amino acids such as tryptophan and phenylalanine. The shikimate pathway is found in plants and microorganisms, never in animals. One important enzyme in the shikimate pathway is 5-enolpyruvyl shikimate 3-phosphase synthase (EPSPS). Glyphosate (Roundup™) is water soluble and can be sprayed upon plant leaves. It is absorbed and then inhibits EPSPS. Glyphosate exhibits little to no toxicity in mammals, birds, and fish because they do not have EPSPS [48]. Roundup Ready® plants carry the gene coding for a glyphosate-insensitive form of this enzyme, obtained from *Agrobacterium* sp. strain CP4. Once incorporated into the plant genome, the gene product, CP4 EPSP synthase, confers crop resistance to glyphosate [49]. This is the basis of so-called "genetically engineered" crops.

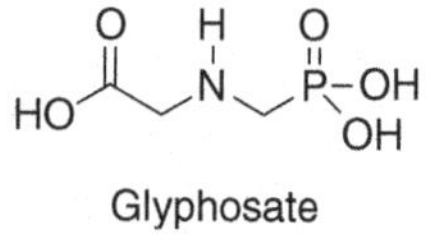

Glyphosate

Glyphosate dominates the herbicide market, in part because of genetically-modified seeds which are resistant to Roundup. Sales data for 2003 shows glyphosate to be \$2.9 billion and the number one ranking active agricultural chemical. Paraquat, the number two ranking agrichemical and also an herbicide is far behind with \$375 million in sales [50]. However, with time some weeds are developing resistance to Roundup, creating a need for herbicides with different modes of action.

Branched-chain amino acids; valine, leucine, and isoleucine are biosynthesized in a pathway catalyzed by the enzyme acetohydroxyacid synthase (AHAS), sometimes referred to as acetolactate synthase (ALS). There are many herbicides that interfere with AHAS. They are classified as group 2 (WSSA) or group B (HRAC).

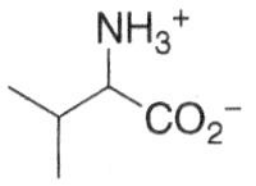

Valine 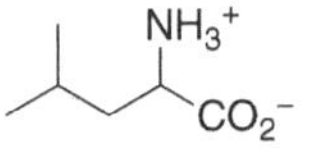 Leucine 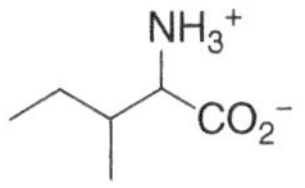 Isoleucine

Certain types of organic compounds inhibit AHAS, resulting in plant death. One reason for the success of AHAS inhibitors is that they are effective at extremely low application rates [51]. The first type discovered was based upon the sulfonylurea moiety. Chlorsulfuron (Glean®, Dupont), a first generation sulfonyl urea was introduced in 1982 [52] for broadleaf control. Second generation sulfonyl ureas replaced the chlorobenzene ring with a substituted pyridine and increased the spectrum of activity. Azimsulfuran is a newer rice herbicide and was first introduced in 1996. About 60% of the global population rely on rice as a major food source [53]. Altogether, more than 30 sulfonylurea herbicides have been commercialized.

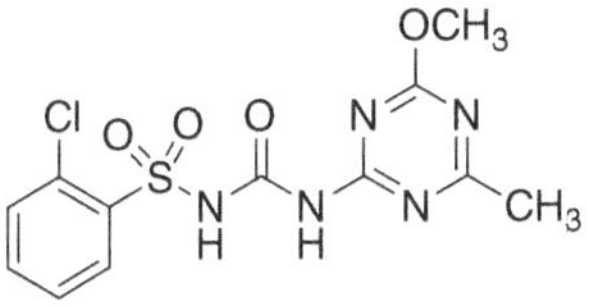

Chlorsulfuran 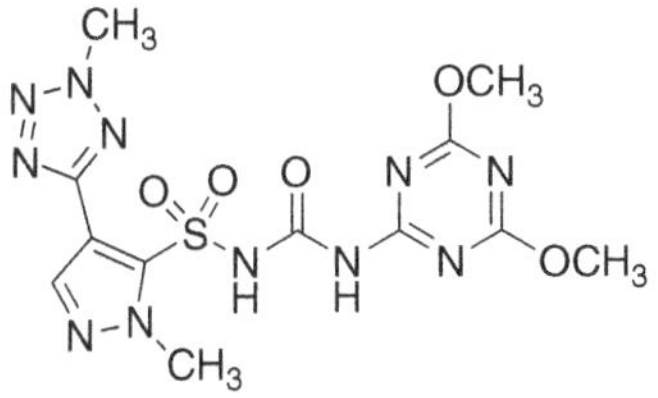 Azimsulfuran

Imidazolinone herbicides are another type of AHAS inhibitors. Imazapic is an example. Other variants either have different substituents on the pyridine ring or substitute a substituted phenyl ring for the pyridine.

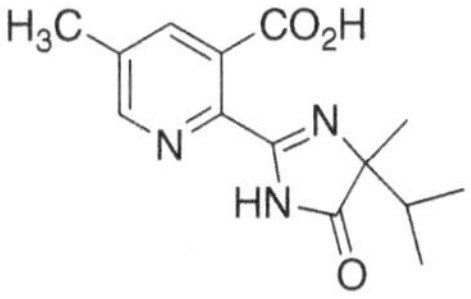

Imazapic

Flumetsulam is one of several in the triazolopyrimidine sulfonamide class of herbicide. This class of AHAS inhibitors is effective for broadleaf weed control. The structure–activity relationship (SAR) has been studied and several variants are commercial. The ideal herbicide is effective at giving a broad spectrum of weed control and not being harmful to the particular crop being grown. Flumetsulam is used for broadleaf weed control in maize and soybeans; metosulam is used for maize and cereals. Diclosulam and cloransulam-methyl are used in soybeans; florasulam is used in cereals [54]. Penoxsulam has a reversed sulfonamide linkage and is used in rice fields to control grass, broadleaf, and sedge weeds [55].

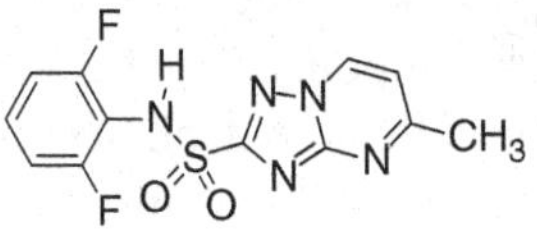

Flumetsulam 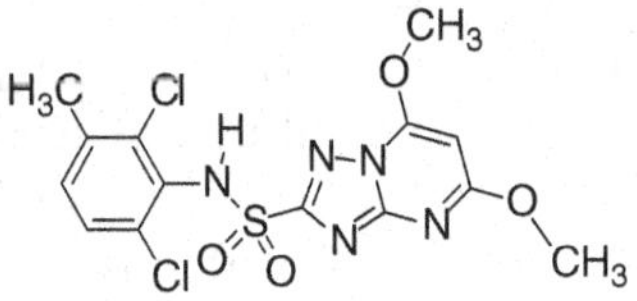 Metosulam 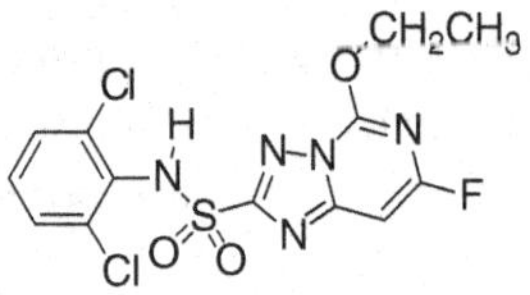 Diclosulam

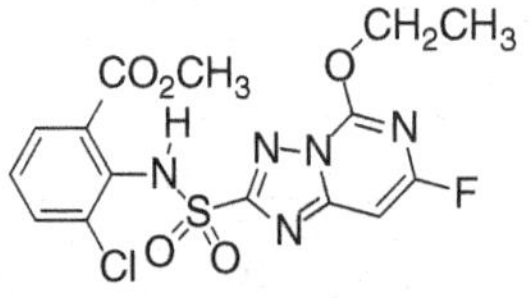

Cloransulam-methyl 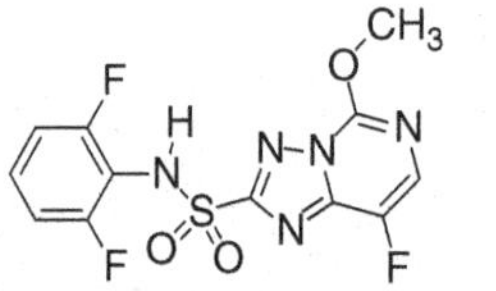 Florasulam 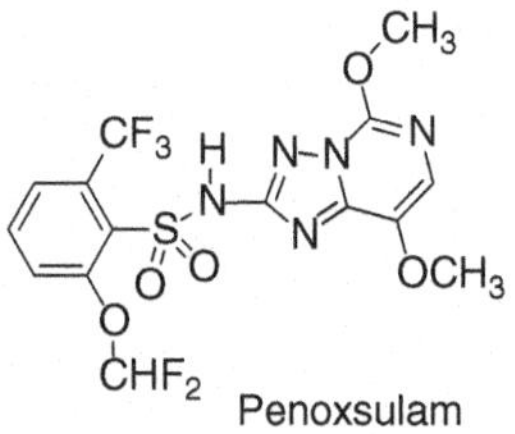 Penoxsulam

One relatively new herbicide is a pyrimidine-based chemical, salflufenacil (Kixor®, BASF) [56]. This herbicide inhibits chlorophyll biosynthesis. In particular it inhibits protoporphyrinogen IX oxidase (PPO) in the tetrapyrrole biosynthetic pathway. Protoporphyrin IX is the heterocylic ring system surrounding the magnesium cation in chlorophyll. There are several herbicides that operate by this mechanism. They are in group 14 (WSSA).

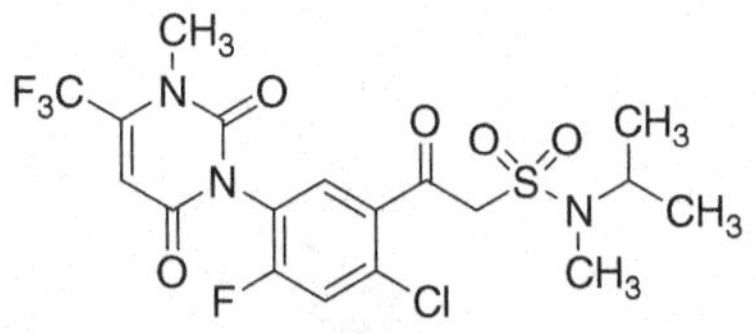

Salflufenacil

One herbicide that has been used since the 1940s is 2,4-dichlorophenoxyacetic acid (2,4-D). It is currently found in about 600 products registered for agricultural, residential, industrial, and aquatic uses [57].

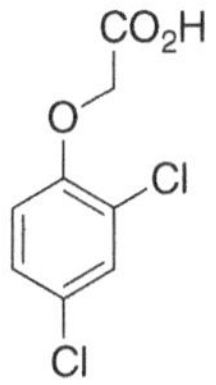

2,4-dichlorophenoxyacetic acid

Chirality can be important in pharmaceuticals and also in other compounds that have biological activity. It can be a factor in herbicides. Dimethenamid (Frontier®, BASF) is an herbicide in group 15, inhibitors of synthesis of very long-chain fatty acids. It can control a variety of annual grasses and several broadleaf weeds and is used for corn crops.

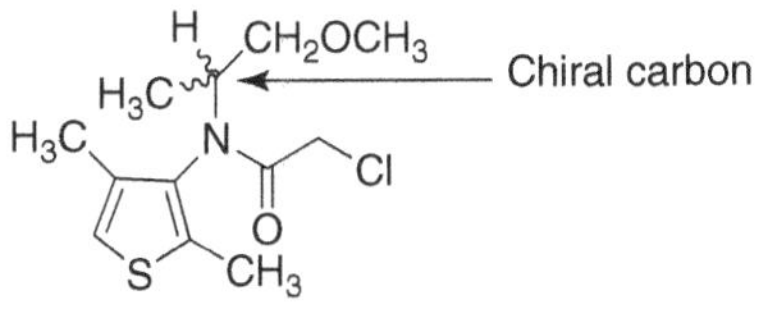

Chiral carbon

Dimethenamid

The molecule has two chiral elements. There is a chiral axis along the bond from the carbon at the 3-position of the thiophene to the nitrogen. This is because there is not free rotation around this single bond. There is also an asymmetric carbon at the methine carbon attached to nitrogen. Because of two chiral elements, there can be 2^2 or four possible stereoisomers. Dimethenamid with S-configuration at the chiral carbon can be prepared from S-methoxyisopropylamine. This single enantiomer amine is isolated from a racemic mixture of amines by enantioselective enzymatic acylation [58]. The enzyme selectively acylates only one enantiomer and the resultant amide can be readily separated from the unreacted free amine. The undesired enantiomer can then be racemized to provide a source of (after further separation) more of the desired enantiomer.

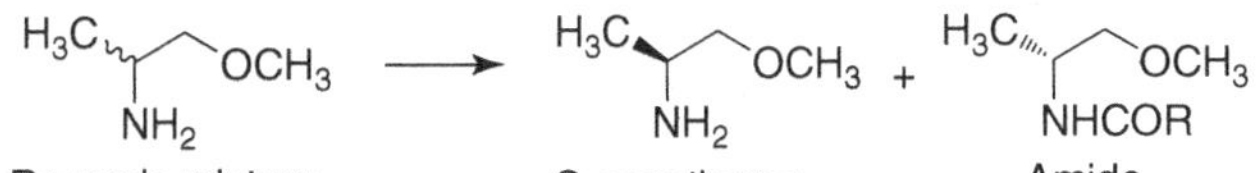

Racemic mixture → S-enantiomer + Amide

Dimethenamid with S configuration at the chiral carbon, (1S,aRS) would be expected to be twice as effective as the previous version with a racemic mixture at the chiral carbon (1RS,aRS) if this were the active diastereomer and the 1R,aRS diastereomer was inactive. However, greenhouse and field tests show it to be three to four times as effective [59]. The 1S,aRS diastereomer (Outlook®, BASF) is marketed for corn and bean crops as a means of controlling grasses and broad-leaf weeds.

Another example of an herbicide where chirality is important is metolachlor (Dual®, Syngenta), also in class 15 and used for corn and soybeans. Metolachlor has two chiral elements. The bond between the nitrogen and the phenyl group is restricted from free rotation and therefore there is a chiral axis. Chiral axes are not particularly common and it is coincidental that they appear in both this and the previous example. Metolachlor also has an asymmetric carbon at the methine carbon attached to nitrogen.

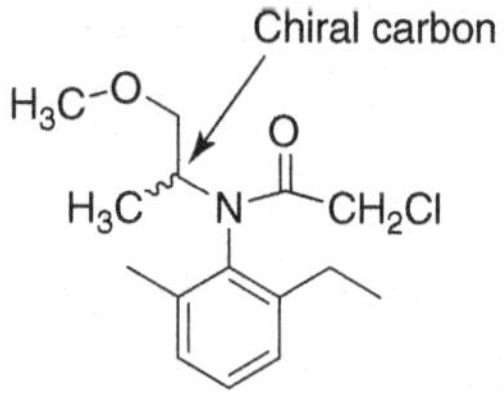

Chiral carbon

Metolachlor

Metolachlor was originally introduced to the market as a mixture of all four stereoisomers. Later it was found that the stereochemistry at the chiral carbon controlled the activity. About 95% of the herbicidal activity comes from the two diastereomers where the methine carbon has S configuration [60]. This recognition led to a lengthy and major industrial research project to prepare metolachlor with S configuration at the methine carbon. The project began in 1981. Fifteen years later, after the evaluation of numerous routes, metal catalysts, chiral ligands, and process conditions catalysts, S-metolachlor was in full-scale production [61]. The key step is the selective hydrogenation of the imine. At greater than 10,000 tons per year, it represents possibly the largest scale enantioselective process. S-Metolachlor is marketed as Dual Gold ®; Dual Magnum® is used as the name in the U.S.

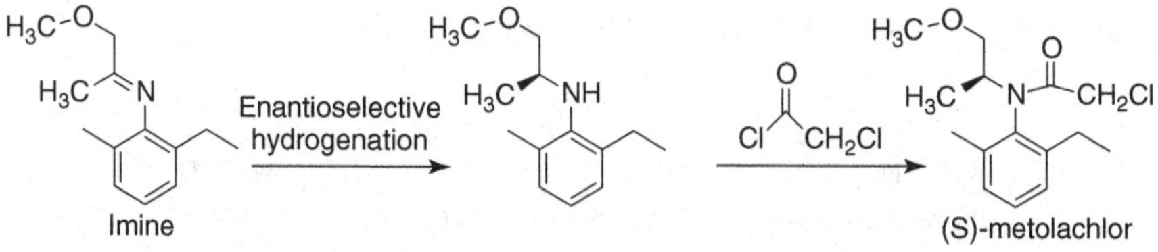

Imine —Enantioselective hydrogenation→ —$ClC(O)CH_2Cl$→ (S)-metolachlor

S-metolachlor is also the active ingredient in Dual II Magnum® (Syngenta). This product also contains a safener, benoxacor. Safener is a name for compounds that protect crops from herbicide injury without replacing herbicide activity in targeted weeds. Various chemicals have been developed to act as safeners for crops such as corn, grains, and rice. Safeners have been used for decades. They increase the activity of detoxification enzymes [62].

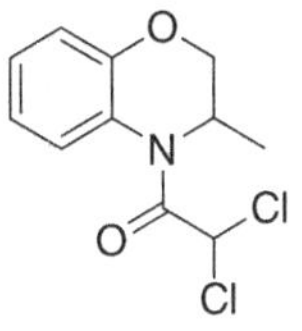

Benoxacor

12.5 FUNGICIDES

Fungicides are used to prevent mold and mildew in crops and seeds. They have been classified by the Fungicide Resistance Action Committee (FRAC) based upon 46 different fungicide and bactericide modes of action. Two of the major groups, based upon percent sales are the DMI fungicides in group G1 and the QoI fungicides in group C3 [63]. The DMI fungicides inhibit sterol biosynthesis. Ergosterol is the predominant sterol and is biosynthesized in several steps involving 11 enzymes from squalene. One of those steps involves removal of a methyl group at the C 14 carbon of the steroid. The group G1 fungicides inhibit this demethylation step and hence are called demethylation inhibitors (DMI) [64].

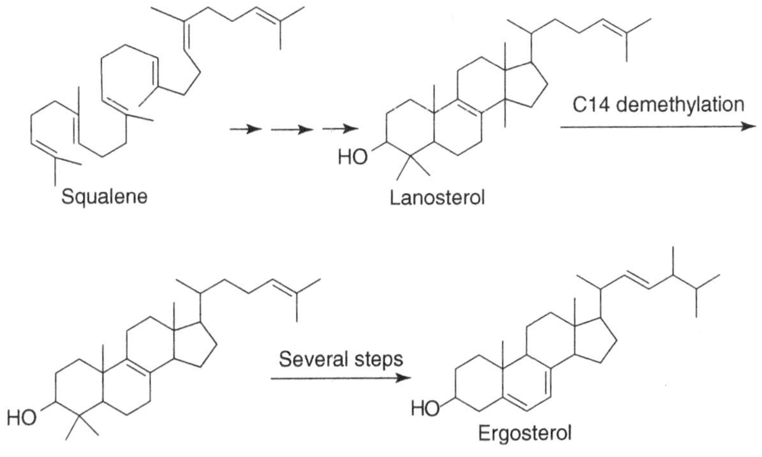

Squalene → → → Lanosterol —C14 demethylation→

—Several steps→ Ergosterol

There are different chemical classes used but DMI fungicides with the triazole ring structure dominate. Examples of DMI fungicides containing the triazole ring include tebuconazole, difenoconazole, and epoxiconazole. The carbon bearing the hydroxy group in tebuconazole is chiral; the herbicide is used as a mixture of enantiomers. Similarly difenoconazole and epoxiconazole have asymmetric carbons and are used as a mixture of stereoisomers. In the case of epoxiconazole, the chlorophenyl group is on the same side of the epoxide as the triazolomethyl substituent.

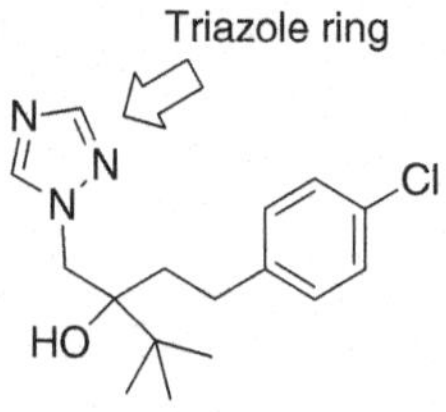

Triazole ring

Tebuconazole 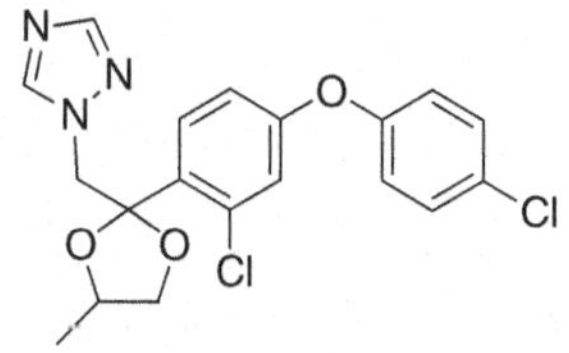 Difenoconazole 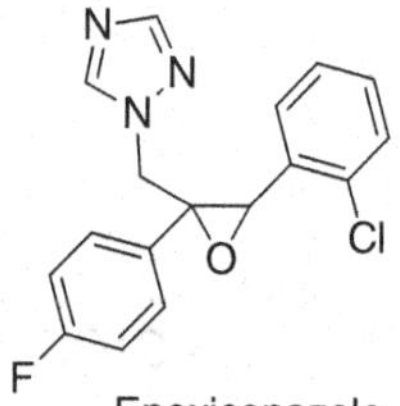 Epoxiconazole

QoI (quinone outside inhibitors) fungicides act at the Quinol outer binding site of the cytochrome bc1 complex thereby inhibiting respiration and causing death of the fungus. Strobilurins such as azoxystrobin and pyraclostrobin dominate this class of fungicides. In 2010, Syngenta sold $1 billion of azoxystrobin under the trade name Amistar [65]. The term strobilurin derives from *Strobilurus tenacellus*, a fungus from which these compounds were originally isolated.

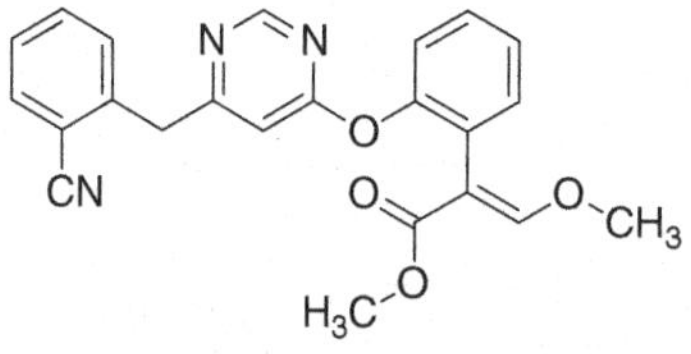

Azoxystrobin 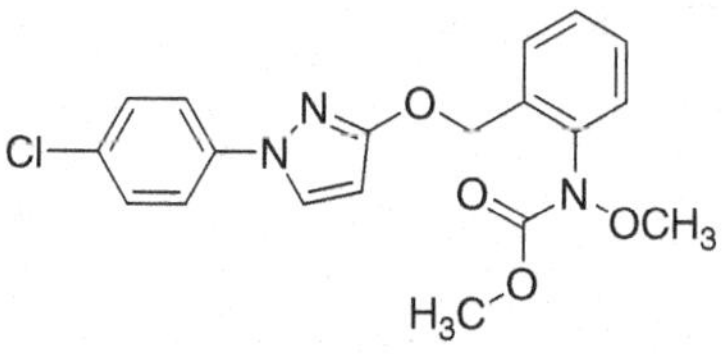 Pyraclostrobin

One class of seed treatment fungicides is the carboxanilides; they are amides of aniline. Carboxin, mepronil, and flutolanil are all commercial members of this class [66]. They are in group C2 and are succinate dehydrogenase inhibitors.

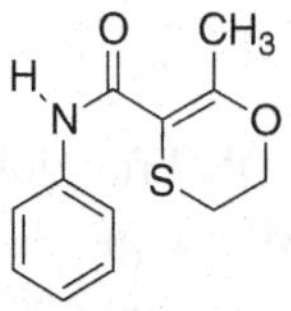

Carboxin 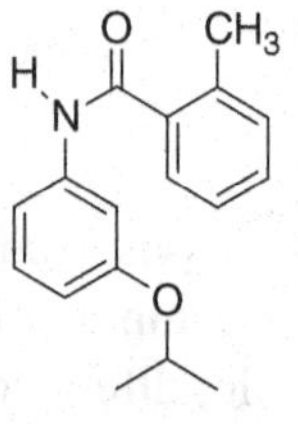 Mepronil 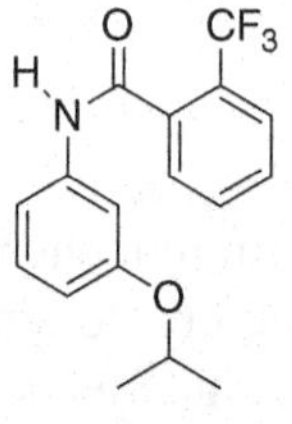 Flutolanil

REFERENCES

1. Hideo Ohkawa, Hisahi Miyagawa, Philip Lee. *Pesticide Chemistry*. New York: Wiley-VCH; 2007:4.
2. Andrew H Cobb, John PH Reade. *Herbicides and Plant Physiology*. 2nd Ed. New York: Wiley-Blackwell; 2010:p. viii.
3. Hideo Ohkawa, Hisahi Miyagawa, Philip Lee. *Pesticide Chemistry*. New York: Wiley-VCH; 2007:44.
4. Britt Erickson. Chemical and Engineering News 2010; 88(13):18.
5. Hideo Ohkawa, Hisahi Miyagawa, Philip Lee. *Pesticide Chemistry*. New York: Wiley-VCH; 2007:10.
6. Melody Bomgardner. Chemical and Engineering News 2011; 89(16):13.
7. Melody Bomgardner. Chemical and Engineering News 2011; 89(16):13.
8. Andrew H Cobb, John PH Reade. *Herbicides and Plant Physiology*. 2nd Ed. New York: Wiley-Blackwell; 2010:p. viii.
9. Ann Thayer. Chemical and Engineering News 2013; 91(21):13.
10. Hideo Ohkawa, Hisahi Miyagawa, Philip Lee. *Pesticide Chemistry*. New York: Wiley-VCH; 2007:23.
11. Y Yao, et al. Environmental Science and Technology 2008, 42:5931–5937.
12. R Dalton, C Boutin. Environmental Toxicology and Chemistry 2010, 29:2304–2315.
13. Langdon Elsworth, Walter Paley. *Fertilizers: Properties, Applications, and Effects*. Nova Science Publishers, Inc., Hauppauge, NY; 2009:p. vii.
14. Langdon Elsworth, Walter Paley, editors. *Fertilizers: Properties, Applications, and Effects*. Nova Science Publishers, Inc., Hauppauge, NY; 2009:30.
15. Langdon Elsworth, Walter Paley, editors. *Fertilizers: Properties, Applications, and Effects*. Nova Science Publishers, Inc., Hauppauge, NY; 2009:34.
16. Langdon Elsworth, Walter Paley, editors. *Fertilizers: Properties, Applications, and Effects*. Nova Science Publishers, Inc., Hauppauge, NY; 2009:29.
17. Langdon Elsworth, Walter Paley, editors. *Fertilizers: Properties, Applications, and Effects*. Nova Science Publishers, Inc., Hauppauge, NY; 2009:34.
18. Langdon Elsworth, Walter Paley, editors. *Fertilizers: Properties, Applications, and Effects*. Nova Science Publishers, Inc., Hauppauge, NY; 2009:38–39.
19. W Kramer, U Schirmer, editors. *Modern Crop Protection Compounds*. Vol. 3. New York: Wiley-VCH; 2007:753.
20. Martins S Odetokun, M Angela Montesano, Gayanga Weerasekera, Ralph D Whitehead Jr., Larry L Needham, Dana Boyd Barr. Journal of Chromatography B 2010; 878(27):2567–2574.
21. Anindita Mitra, Chandranath Chatterjee, Fatik B Mandal. Research Journal of Environmental Toxicology 2011; 5:81–96.
22. Renata Raina, Lina Sun. Journal of Environmental Science and Health Part B 2008; 43:323–332.

23. Matthew R Bonner, Joseph Coble, Aaron Blair, Laura E Beane Freeman, Jane A Hoppin, Dale P Sandler, Michael C R Alavanja Am. J. Epidemiol. 2007; 166:1023–1034.
24. Michael McCoy. Chemical and Engineering News 2011; 89(11):13.
25. Anindita Mitra, Chandranath Chatterjee, Fatik B Mandal. Research Journal of Environmental Toxicology 2011; 5:81–96.
26. Irma Kakko, Tarja Toimela, Hanna Tähti. Environmental Toxicology and Pharmacology 2004; 15(2–3):95–102.
27. Kazuhiko Matsudaa, Yukio Kikutaa, Atsushi Habaa, Koji Nakayamab, Yoshio Katsudab, Akikazu Hatanakac, Koichiro Komai. Phytochemistry 2005; 66(13):1529–1535.
28. Leslie Crombie. Pestic. Sci. 1980; 11:102–118.
29. David Pulman. J Agric. Food Chem. 2011; 59:2770–2772.
30. Virginia Perez-Fernandez, Maria Angeles Garcia, Maria Luisa Marina. Journal of Chromatography A 2010; 1217:968–989.
31. Hideo Ohkawa, Hisahi Miyagawa, Philip Lee. *Pesticide Chemistry*. New York: Wiley-VCH; 2007:149–150.
32. W Kramer, U Schirmer, editors. *Modern Crop Protection Compounds*. Vol. 3, New York: Wiley-VCH 2007:767.
33. Motohiro Tomizawa, Todd T Talley, David Maltby, Kathleen A Durkin, Katalin F Medzihradszky, Alma L Burlingame, Palmer Taylor, John E Casida, PNAS 2007; 104:9075–9080.
34. Motohiro Tomizawa, John E Casida. Accounts of Chemical Research 2009; 42:260–269.
35. Hideo Ohkawa, Hisahi Miyagawa, Philip Lee. *Pesticide Chemistry*. New York: Wiley-VCH; 2007:261–262, 271.
36. Glenn Hess. Chemical and Engineering News 2012; 90(31):36.
37. Elizabeth Wilson. Chemical and Engineering News 2012; 90(14):10.
38. Sydney Camerona, et al. PNAS 2011; 108(2):662–667.
39. Britt Erickson. Chemical and Engineering News 2013; 91(18):11.
40. Britt Erickson. Chemical and Engineering News 2013; 91(34):20.
41. W Kramer, U Schirmer, editors. *Modern Crop Protection Compounds*. Vol. 3. New York: Wiley-VCH; 2007:1049.
42. Britt Erickson. Chemical and Engineering News 2013; 91(29):21.
43. Vincent Corbel, Maria Stankiewicz, Cédric Pennetier, Didier Fournier, Jure Stojan, Emmanuelle Girard, Mitko Dimitrov, Jordi Molgó, Jean-Marc Hougard, Bruno Lapied. BMC Biology 2009; 7:47.
44. Gretchen Paluch, Lyric Bartholomay, Joel Coats. Pest Manag. Sci. 2010; 66:925–935.
45. W Kramer, U Schirmer, editors. *Modern Crop Protection Compounds*. Vol. 3. New York: Wiley-VCH 2007:141,145.

46. Andrew H Cobb, John PH Reade. *Herbicides and Plant Physiology*. 2nd Ed. New York: Wiley-Blackwell; 2010:1.
47. Carol Mallory-Smith, E James Retzinger. Weed Technology 2003; 17:605–619.
48. W Kramer, U Schirmer, editors. *Modern Crop Protection Compounds*. Vol. 1. New York: Wiley-VCH; 2007:291.
49. Todd Funke, Huijong Han, Martha L Healy-Fried, Markus Fischer, Ernst Schönbrunn. PNAS 2006; 103(35):13010–13015.
50. Andrew H Cobb, John PH Reade. *Herbicides and Plant Physiology*. 2nd Ed. New York: Wiley-Blackwell; 2010:31.
51. W Kramer, U Schirmer, editors. *Modern Crop Protection Compounds*. Vol. 1. New York: Wiley-VCH; 2007:3.
52. W Kramer, U Schirmer, editors. *Modern Crop Protection Compounds*. Vol. 1. New York: Wiley-VCH; 2007:32.
53. W Kramer, U Schirmer, editors. *Modern Crop Protection Compounds*. Vol. 1. New York: Wiley-VCH; 2007:63.
54. Hideo Ohkawa, Hisahi Miyagawa, Philip Lee. *Pesticide Chemistry*. New York: Wiley-VCH; 2007:89.
55. Hideo Ohkawa, Hisahi Miyagawa, Philip Lee. *Pesticide Chemistry*. New York: Wiley-VCH; 2007:99.
56. Melody Bomgardner. Chemical and Engineering News 2011; 89(16):14.
57. Britt Erickson. Chemical and Engineering News 2012; 90(16):33.
58. A Maureen Rouhi. Chemical and Engineering News 2004; 82(24):62.
59. Karl Seckinger, et al. U.S. Pat. No. 5,457,085. 1995.
60. H Blaser, et al. Acc. Chem. Res. 2007; 40:1240–1250.
61. H Blaser. Adv. Synth. Catal. 2002; 344(1):17–31.
62. D Riechers, K Kreuz, Q Zhang. Plant Physiology 2010; 153:3–13.
63. W Kramer, U Schirmer, editors. *Modern Crop Protection Compounds*. Vol. 2. New York: Wiley-VCH 2007:417.
64. W Kramer, U Schirmer, editors. *Modern Crop Protection Compounds*. Vol. 2. New York: Wiley-VCH 2007:609.
65. Melody Bomgardner. Chemical and Engineering News 2011; 89(16):16.
66. Hideo Ohkawa, Hisahi Miyagawa, Philip Lee. *Pesticide Chemistry*. New York: Wiley-VCH; 2007:295–296.

CHAPTER 13

Design of Experiments and Statistical Process Control

13.1 INTRODUCTION

The topics of this chapter involve a certain amount of mathematics and many associated texts are full of mind-numbing pages of equations. The object here is to introduce the reader to the subjects, explain their importance and give some sense of how they are used. Some use of mathematics is necessary, but every attempt will be made to keep the examples simple so we can keep focused on the larger picture of why these techniques are useful.

In the discussion of pharmaceuticals and the importance of standard operating procedures, an example involving the flying of paper airplanes was used. Curious as to my talent in such an endeavor, I armed myself with several sheets of white copy paper and went to an empty classroom on the first floor. I assembled and flew four paper airplanes and each time carefully measured the flight distance. They flew 12.7, 13.1, 12.5, and 13.0 feet. A bar chart (Figure 13.1) can be easily made.

The average or mean can be calculated by the sum of the individual measurements divided by the number of measurements. In this case, the four flights totaled 51.3 feet, which upon dividing by 4 gives 12.825 feet as the average. Because each of the measurements was done to a precision of three significant figures, strictly speaking the average should also be reported to the same level of precision, namely 12.8 feet. Note that the number "4" in this example is an exact number (there were exactly 4 flights) so this does not imply a precision of one significant figure.

We can get some idea of the variability of these numbers by calculating the standard deviation. The standard deviation is the square root of the variance. The variance is the sum of the squares of the difference of each value from the mean divided by the number of samples. By squaring the numbers, the sign

Fundamentals of Industrial Chemistry: Pharmaceuticals, Polymers, and Business, First Edition. John A. Tyrell.

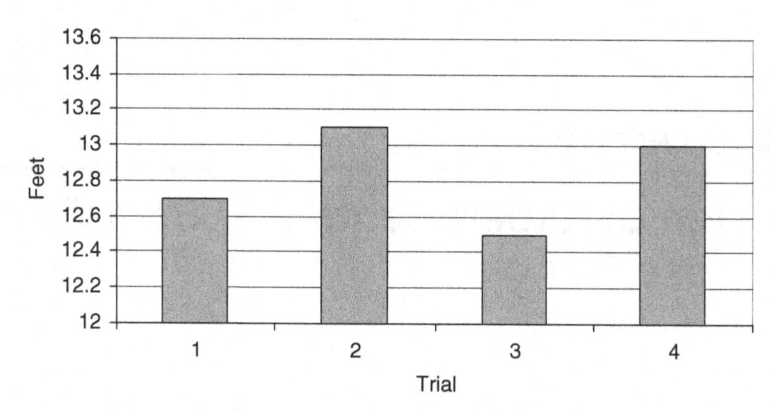

Figure 13.1 White Paper Airplane Flights

is removed and differences above and below the mean are treated equally. In this example, the variance =

$$(1/4)((12.7 - 12.8)^2 + (13.1 - 12.8)^2 + (12.5 - 12.8)^2 + (13.0 - 12.8)^2)$$
$$= 0.057$$

The standard deviation is the square root of 0.057 = 0.24 feet.

Sometimes the sample standard deviation is used when the numbers represent a sampling. In that instance, the sample standard deviation is calculated by taking the square root of the sum of the squares of each of the individual variances divided by one less than the number of samples. In the above example, we would divide by three (4–1) rather than four. This removes a bias so it is sometimes referred to as the unbiased standard deviation. Unfortunately, the term "standard deviation" is used to refer to both versions. To keep things simple, I will not use the sample standard deviation in the examples, but you should be aware of it because it might be used in other texts.

The smaller the variance and the standard deviation, the closer the numbers are to each other. In the example, my flights were similar and thus we have a low variance and a low standard deviation. If we have large variances, sometimes that is an indication that something is wrong with the test. If the data fits a bell-shaped normal distribution curve, 68% of the numbers will fall within one standard deviation of the mean, 95% within two standard deviations, 99.7% within three standard deviations, and 99.99% within four standard deviations. Sometimes charts use error bars to indicate the reliability of the numbers. The error can be calculated in different ways, but one common technique is to use plus or minus one standard deviation. Done this way, my chart (Figure 13.2) would look like this.

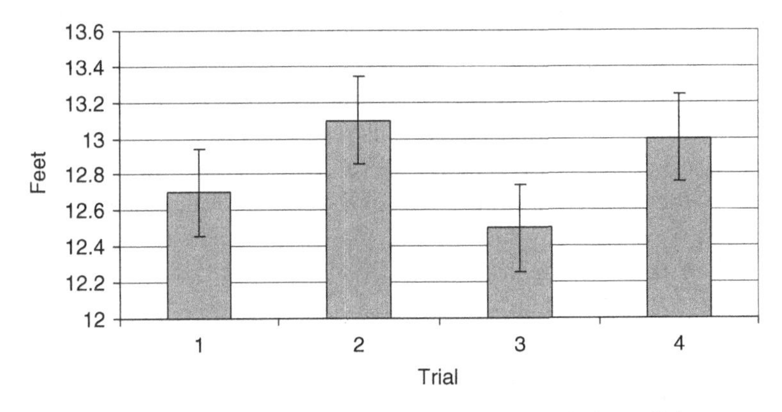

Figure 13.2 Standard Deviation of White Paper Airplane Flights

I began thinking that my prowess at paper airplanes wasn't quite what I had imagined and decided to give it another try. I was out of my white copy paper but was able to borrow several sheets of blue paper from a colleague. The downstairs classroom was occupied, but I was able to find an empty classroom on the second floor. I made and flew four blue paper airplanes and measured the distance flown. They flew 16.8, 17.1, 17.5, and 16.6 feet. We can calculate that the mean is 17.0 feet and the standard deviation is 0.68 feet. The results from the four white airplanes with a mean of 12.8 feet can be graphically compared (Figure 13.3).

It seems that the blue airplanes fly further than the white airplanes. There are statistical tests that can be used to determine whether the blue airplane results differ from the white airplane results in a significant way. Note that in this context, the term "significant" means statistically significant and is not the same as "important." For example, a change in a polymer process may result in a statistically significant difference in polymer tensile elongation,

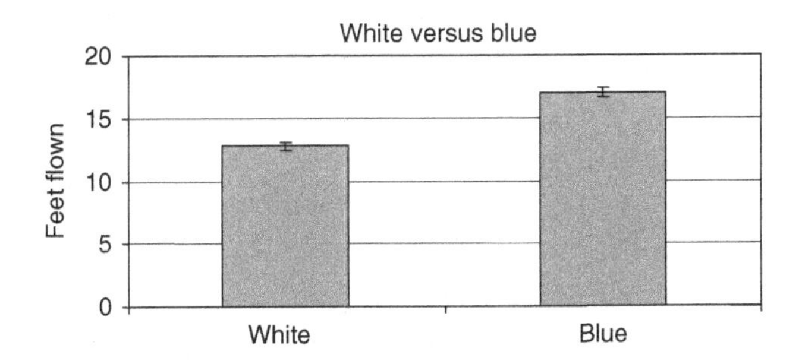

Figure 13.3 White versus Blue Paper Airplane Flights

but the difference in tensile elongation may not be important or a significant property for many applications.

One such method of determining significance is to perform a paired t-test, also known as a two-sample t-test. This test assumes that the data are independent and normally distributed. The value for t is the difference in means divided by the square root of the sum of the square of the standard deviation of the first set divided by the number of tests and the sum of the square of the standard deviation of the first set divided by the number of tests. In the example,

$$t = \frac{17 - 12.8}{[(0.68)^{2/4} + (0.48)^{2/4}]^{1/2}} = \frac{4.2}{[0.173]^{1/2}} = 10.1$$

Tables of t values can be consulted to give the probability that there is a difference between the two data sets. In this example, if the t value is greater than 5.96 we can say that there is a greater than 99.9% probability that there is a difference between the two data sets. Because the t value is 10.1 we can say that there is greater than 99.9 % probability that the data for the blue airplanes is different from the white airplane data. Another way to express this is to say that there is a significant difference.

Continuing with the example, I decided to see how red paper worked. At this time, all the classrooms were occupied so I conducted my test flights outdoors. I made and flew four red paper airplanes and measured the distance flown. They flew 18.8, 19.1, 18.5, and 19.6 feet. We can calculate that the mean is 19.0 feet and the standard deviation is 0.41 feet (Figure 13.4).

The data indicates that red airplanes fly better than blue ones which in turn fly better than white ones. How can this be true? Should we contact Boeing and NASA about our discovery? The answer is that it is not true. The color

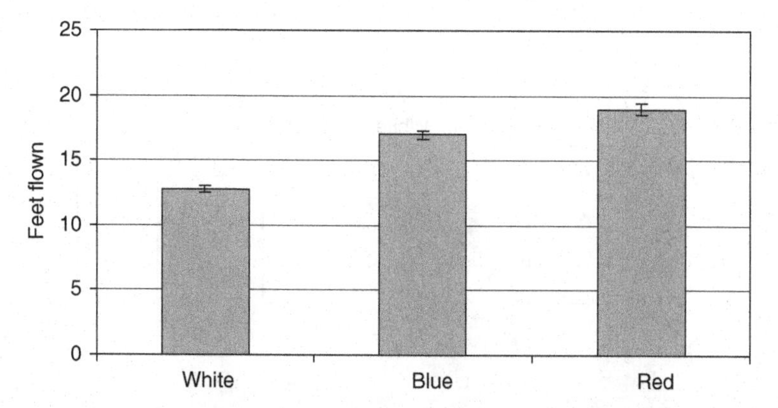

Figure 13.4 White versus Blue versus Red Paper Airplane Flights

of the paper correlates with the length of the flight but it does not cause it. What is more likely is that some other variable caused the differences. Perhaps the airflow in the two classrooms and outdoors had an effect. A likely explanation is that I was simply getting better at making and flying paper airplanes. Because I tested the red ones last, those are the ones that gave the best results. This was a simple illustration of correlation versus causation. Another example is the rooster. Each day the rooster crows and shortly thereafter the sun rises. The crowing of the rooster correlates with the dawn but no one, except perhaps the rooster, thinks that it is the crowing that causes the sun to rise.

These examples seem obvious and silly but mistakes like this happen all the time. They are just not so obvious. Imagine a scientist wishing to know the effect of a certain variable such as temperature, agitation, solvent, or time on reaction yield. Typically they will hold all other recognized variables as close to constant as they can and try different levels of one variable such as agitation to measure the effect. If the yield at low, medium, and high agitation is 45%, 52%, and 57% respectively, they may conclude that increased agitation improves yield. Just as we did with the paper airplanes, they back up the conclusion with graphs and statistics. However, this may be a similar situation to the paper airplanes; they are just becoming more proficient at performing the reaction and agitation does not have an effect on yield. Perhaps the effect is caused by some other unrecognized hidden variable such as humidity level in the laboratory when the reaction was performed and has nothing to do with agitation. With this approach, the scientist may never recognize this.

In another example, if the yield at low, medium, and high agitation is 45%, 47%, and 44% respectively, one may conclude that the amount of agitation is not important. This may be correct for the temperature, solvent and time that was used. But we can imagine that perhaps agitation is important with a different solvent system. This effect may never be recognized when changing variables one at a time.

Consider the contour chart of yield with temperature on the y axis and time on the x axis (Figure 13.5). The yield varies from less than 10% to about 50%.

If we were to run experiments at a fixed time of two hours and vary the temperature from 10° to 50° in 10° increments, our results might be (Figure 13.6):

And we would conclude that a temperature of about 20°C and a yield of about 22% is optimal. However, in this example, temperature and time are interacting variables. Notice that with this approach, we would never find the optimal yield of about 50% which can be achieved at a temperature near 35°C and a time of about 4 hours.

Interacting variables is just one problem that can cause someone to draw the wrong conclusion. Another is random variation. Imagine that a reaction

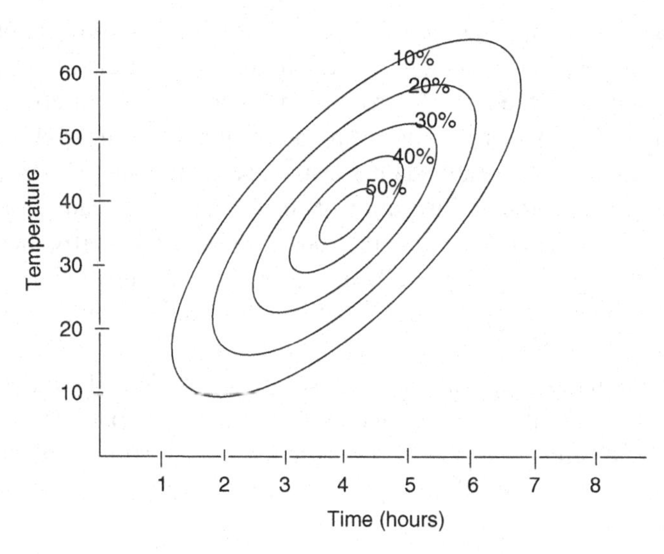

Figure 13.5 Contour Chart of Yield versus Temperature and Time

Temperature	Yield
10°	10%
20°	22%
30°	19%
40°	8%
50°	2%

Figure 13.6 Yield versus Temperature

run under the same conditions gives yields between 80% and 90%. An impatient experimentalist might check the importance of agitation and perform two experiments, one at low agitation, obtaining a yield of 90%, and a second experiment at high agitation, obtaining a yield of 80%. It would be logical to think that agitation is important and that it is better to use low agitation. However, in this example, agitation has no effect and the impatient experimentalist was just observing random variation.

Another potential pitfall is confounding factors. Imagine a reaction where time is very important and temperature does not have much effect. A set of experiments to determine the effect of temperature might give increasing yields as temperature is increased, but the reason might be that experiments take longer to run at higher temperatures due to the time to heat the reaction mixture and the observed effect is due to the time differences not the temperature effects. In this example, time is a confounding factor or a confounding variable.

One approach to avoid these pitfalls is to use statistical experimental design.

13.2 DESIGN OF EXPERIMENTS

Design of experiments, also called experimental design, is a plan to study the various variables and their effect on the desired result, often yield or purity. If planned and interpreted correctly, interactions and hidden variables can be uncovered. An experimental design should include all relevant variables and there should be enough experiments performed to determine important effects and eliminate trivial variables.

Consider the situation of Sam. He is a junior process chemist associated with a large volume process. Sam thinks that there may be a process advantage obtained by increasing the agitation. He goes to the lab and runs two experiments. The one with the current level of agitation gives a yield of 65%. He then increases the stirring rate in the second experiment and obtains a yield of 76%. He writes in his monthly progress report that he has obtained a yield improvement of about 10%. The report is circulated throughout the business. Everyone is excited about this breakthrough because a 10% yield improvement will have a huge impact on the profitability of the business. A team is assembled to pursue this lead. They run three more experiments with low agitation and three with high. Including the original data, the percent yields with low agitation are: 65, 72, 67, and 75 for an average of 69.75%. With high agitation, the yields are 76, 65, 68, and 69 for an average of 69.5%. Comparing the averages, there is no effect of agitation on yield. The team reports this result. What do you think has happened to Sam's status and credibility within the company?

Consider a second situation with Josephine at another company. She too is working on a large volume process and does an experiment with the current level of agitation, obtaining a yield of 65%, and the one with high agitation, obtaining a yield of 66%. She reports that there is no effect of agitation on yield. Had she done a total of four experiments for each condition, she would see percent yields of 65, 64, 62, and 62 for the standard agitation and 66, 71, 69, and 72 for high agitation. The average yield for high agitation is 69.5% versus 63.25% for standard agitation. By doing the comparison of the single runs, Josephine missed seeing the real effect and the potential to improve yield by varying a simple process variable. Had she compared averages, she would not have missed this.

The previous two examples are plausible and should serve to convince you of the value of comparing averages versus single data points. The number of individual runs that should be averaged depends upon the reproducibility of experiments. The greater the precision, the fewer runs need to be averaged. It also depends upon the magnitude of change that is of interest. The smaller the magnitude of change that is of interest, the greater the number of experiments that should be averaged. Commonly 4 to 16 experiments are averaged.

Let us consider a situation where we were not interested in minor effects and had fairly good experimental precision. We might be able to compare the effect of a variable by comparing the average of four experimental results for each setting. So to look at agitation we might run four experiments with high agitation and four with low and compare the averages.

If we wanted to look at another variable, for example, the effect of temperature, we would need another eight experiments. To look at both variables, we have now run 16 experiments. Often an experiment can be long and tedious. To run 16 experiments might require more than a month of laboratory time. If we combine the experiments and vary both the agitation and temperature, we can "double dip", essentially learning about two variables in eight experiments. This is called a two-factor design. The experimental design would look like this (Figure 13.7):

It is important that the experiments be run in a random order. The eight experiments can be used to see the effect of either variable. We have saved ourselves eight experiments by having these experiments perform double duty. If agitation and temperature interact with each other we would not see the effect if we studied each variable independently, but we would see the effect with

Experiment number	Agitation	Temperature
1	High	High
2	High	High
3	High	Low
4	High	Low
5	Low	High
6	Low	High
7	Low	Low
8	Low	Low

Figure 13.7 Two Factor Experimental Design

this design. By planning the experiments, we are able to do fewer experiments and gain more information.

Consider an example with three variables, e.g. perhaps in a synthesis, time, temperature, and catalyst loading, would be important variables. The design of experiments would vary these three variables and be called a three-factor design or a three-way design. Often they are varied at two levels, the practical extremes. Perhaps the temperature would be studied at 20°C and 90°C, the time at 1 hour and 6 hours and the catalyst loading at 50 ppm and 500 ppm. Each of these extremes would be chosen by the experimenter as what makes sense for the reaction. Because two levels are being studied, this design is a three-factor design with two levels. Completion of a full factorial design would require all of the combinations to be tested. For a three-factor design with two levels, there are $2^3 = 8$ combinations and this might be called a 2^3 full factorial design. Analysis of the data allows the experimenter to understand interactions and main effects. The experiments should be done in random order to avoid being misled by any unknown variables such as improving proficiency in the synthesis or deteriorating purity of one of the raw materials. It is also helpful to include one or two replicates and perhaps a center point such as medium temperature, time, and catalyst loading. If you plan ten experiments, labeling them from one to ten, you can simply write the ten numbers on slips of paper, shake them in a jar or hat and pull them out; performing the experiments in the order you pick them from the jar.

There are software packages available that will help design the experiments, give an order for their execution, help with the data interpretation, and print graphs with the results. As with any software, it is important for the scientist to avoid abdication of all reasoning to the computer. It is important for the scientist to think about the experiments and the results and make sure they are consistent with scientific principles. Early in his career, the author consulted with the company statistician who employing software guided the experimental design for the reaction of bromine with acetone to synthesize 1,1,3- tribromoacetone. After the design was performed, the statistician consulted the computer analysis of the results and declared that bromine was not necessary. Any chemist recognizes that this is wrong. The effect was parabolic and our levels were at opposite sides of the parabola. This confusion would have been eliminated by choosing to do a center point experiment, but nonetheless illustrates the importance of thinking about the science behind any observations.

Often there are many more than three important variables. Even with only five variables, we are at $2^5 = 32$ experiments for a full factorial design. We would prefer to also have a center point and one or two replicates, so now we are at 35. Dependent upon the difficulty in performing each experiment, this can be a large undertaking. One option is to undertake a fractional factorial design, and perform a fraction of the experiments. A half-factorial design of a five–factor design would involve 16 experiments, half of the 32 needed for a full-factorial design. With a half-factorial design, the main effects and first-order interactions can still be determined. If there is any ambiguity, further experiments can be subsequently performed. In 2011, as part of their Quality by Design initiative, the FDA encouraged design of experiment techniques in new drug applications with the goal of improving the quality of drugs and ensuring supply-chain safety[1].

One variation is called Evolutionary Operation of Processes (EVOP). This technique involves studying small changes to a process during normal operation, to find the optimum operating conditions. The advantage is that it can use data already being generated by an existing process and has minimal disruption on manufacturing.

13.3 STATISTICAL PROCESS CONTROL

Processes must be controlled for quality purposes. Most major chemical processes have automatic computerized data logging and many parameters are automatically recorded and available for study. These can include set points, process measurements such as flow rate, temperature, pressure, stirring speed, and so forth or product parameters such as composition, density, clarity, and so forth, especially when the product is measured on-line.

Let's consider a simple process to make a formulation containing 50% by weight of component A. Component A and the other ingredients are continuously fed from different belts into a mixer. The blend continuously exits the mixer and is measured for the weight percent component A in the blend. For the sake of this example, let's assume that the measurement is made once a shift or once every eight hours. The process operators have the option to adjust the feed belt speed and therefore the amount of each component fed to the mixer. If the blend goes out of quality, they are expected to make these adjustments so that the product is suitable for sale. The process has been running fine. Imagine that it runs for almost another week with no process adjustments and with the following results (Figure 13.8):

Had you been the operator during this run, would you make any adjustments? After the first measurement of 48%, which is lower than the desired 50%, it might be tempting to adjust the belt speeds to increase the feed of component A. However, with no adjustment, the second result was 51.3%. With an adjustment it likely would have been much higher and therefore further from the target of 50%. How did the result go from 48% to 51.3%? Is it significant? Is it reflective of a need for a process change such as a change in the feed belt speed or is it normal process variation. Or is it variability in the test measurement? Or a combination?

Sample Number	Result
1	48
2	51.3
3	49.5
4	52.3
5	50.6
6	47.3
7	50.2
8	51.1
9	50
10	50.7
11	49.6
12	49.7
13	48.2
14	50.3
15	46.5
16	49.9
17	53.3
18	52.4
19	48.5
20	52.8

Figure 13.8 Table of Process Results

We can graph the results (Figure 13.9):

It is still not clear. If we calculate the average and the standard deviation and assume a normal distribution, we can say that there is a 99.7% likelihood that a number will fall within three standard deviations. The average is 50.1% and the standard deviation is 1.8. This tells us that if it is a normal distribution, there is a 99.7% probability that a random point will fall between 44.7% (50.1 − 3(1.8)) and 55.5% (50.1 + 3(1.8)). Replotting with lines for three standard deviations, our plot looks like this (Figure 13.10).

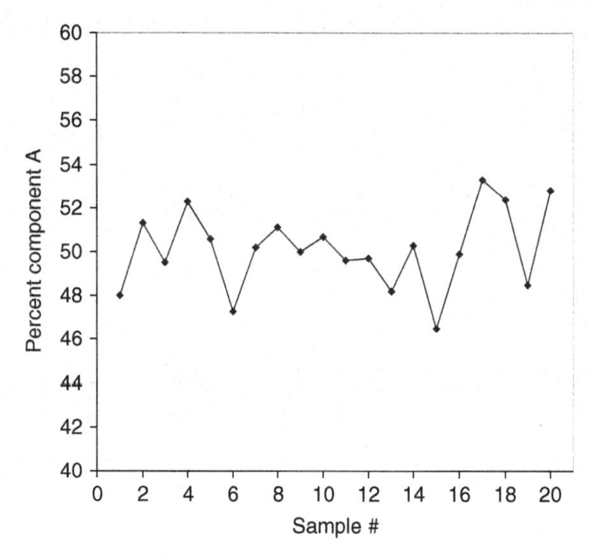

Figure 13.9 Graph of Process Results

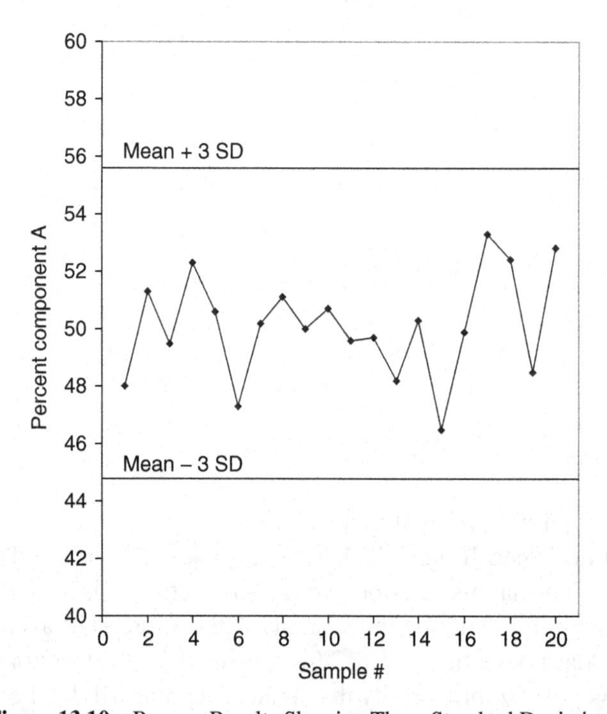

Figure 13.10 Process Results Showing Three Standard Deviations

Because all of the points are within three standard deviations, we expect that we are just seeing normal process deviations. If we plot historical values for the process, we should see that they typically fall within upper and lower limits similar to the upper and lower limits calculated by taking the mean ± 3 standard deviations. We can use this charting method to use statistical process control. We do this by setting guidelines for control of the process. One guideline might be that no adjustments should be made if the process is "in control". We can define "in control" as no sample points outside of the limits (can use either the process limits from historical data or the calculated limits); most points near average; similar number of points above andbelow average; and a random distribution of points.

If we accept these control limits and continue to run the process without adjustment, checking again after 10 more shifts, we see the control chart (Figure 13.11). The process is out of control. Beginning with sample 22,

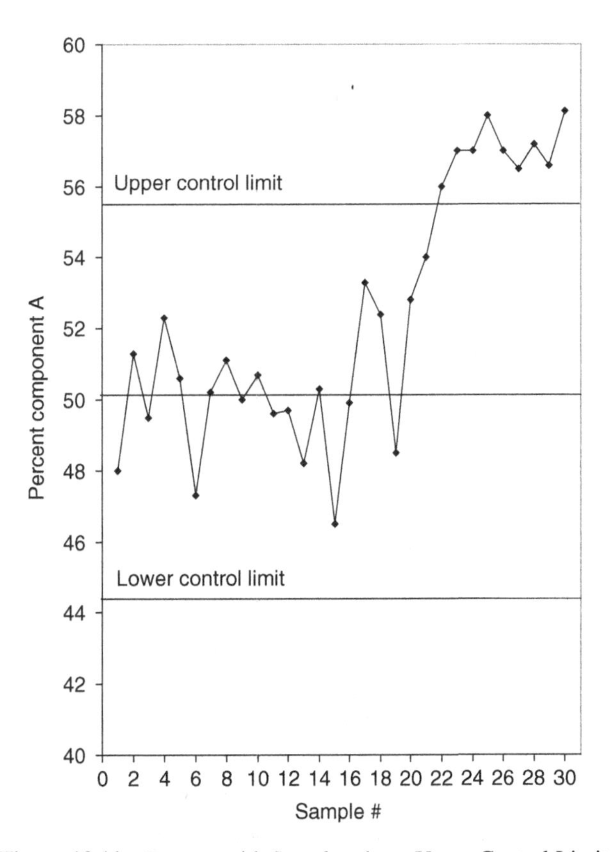

Figure 13.11 Process with Samples above Upper Control Limit

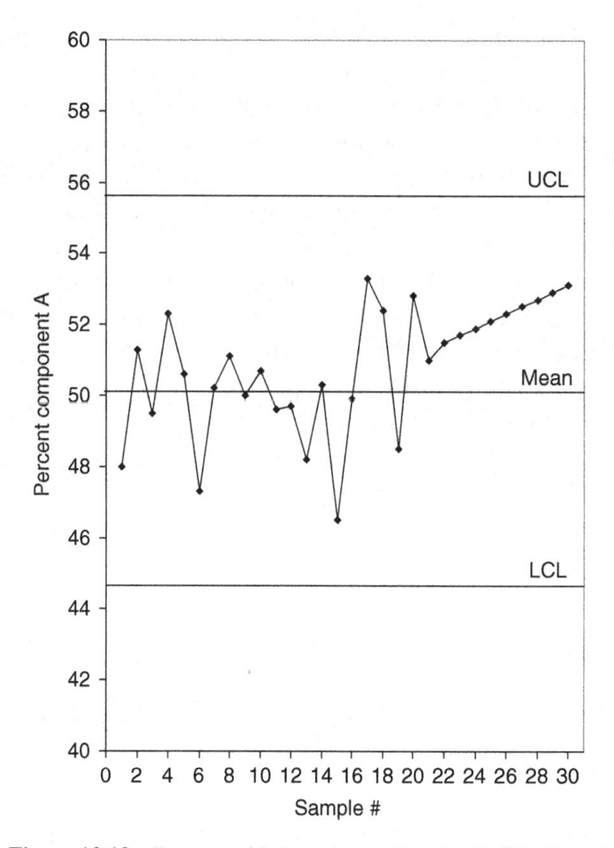

Figure 13.12 Process with Samples not Randomly Distributed

we are above the upper control limit. That shift should have taken corrective action and made adjustments.

What if the next ten samples had looked like the next chart (Figure 13.12)? All the data points are within the control limits, but the process is not in control. The points are not randomly distributed, and there are not a similar number of points above and below the mean. Corrective action should be taken. This also began around shift 22 but the trend would probably not be recognized until several more samples had been tested.

One of the advantages of control charting and statistical process control is that everyone seems to understand the use and it can be a tool to improve process consistency. With a set of rules, it can be used to prevent excessive "tweaking" of a process which can cause increased variability and therefore decreased quality. It can also give increased process understanding. In the previous two examples, something changed around shift 22. By going back through the data logs, the process engineer can often gain an understanding about what the change may have been and take actions to prevent its reoccurrence. This enables continuous process improvement.

QUESTIONS

1. (Mark **each** of a-d as True or False) Design of Experiments ...
   a. can prevent confusion of correlation with causation
   b. can prevent confusion due to experimental error
   c. is more efficient than application of scientific principles
   d. cannot be used if there are interacting variables
2. Examine the process data below. The data is listed sequentially and represents 20 samples taken every eight hours and the resultant specific gravity results.
   0.71, 0.69, 0.72, 0.69, 0.69, 0.71, 0.71, 0.69, 0.76, 0.76, 0.77, 0.76, 0.77, 0.78, 0.76, 0.77, 0.69, 0.69, 0.70, 0.69
   Which is true?
   a. The process is "in-control"
   b. The process is "out of control"
   c. We do not know if the process is in control because the problem does not tell us the process limits.
   d. We do not know if the process is in control because the problem does not tell us the type of synthesis.
3. Explain your selection for problem 2.

REFERENCE

1. Rick Mullin. Chemical and Engineering News 2013; 91(13):27.

CHAPTER 14

Safety and Environmental Considerations

14.1 SAFETY AND RESPONSIBLE CARE

There is an old saying, "*If you think safety is expensive, try having an accident.*" The safe production, transport, and use of chemicals require that certain precautions be taken. There have been some terrible tragedies involving chemicals. In April 1947, a fire aboard a ship docked at Texas City, TX caused the explosion of a cargo of thousands of tons of ammonium nitrate. More than 500 people died and more than 3,000 were injured. In 1984, at a Union Carbide plant in Bhopal, India, there was a release of methyl isocyanate. The plant was in a highly populated area and more than 3,000 people died. In 1989, there was a fire and explosion at a Phillips polyethylene plant in Pasadena, TX, resulting in the death of 23 people. The accident happened when flammable process gases were released during a maintenance procedure. In 2013, 66 years after the Texas City explosion, another ammonium nitrate explosion occurred at the West Fertilizer Co. in Texas, killing more than a dozen people.

In response to the Bhopal tragedy, the U.S. enacted the Emergency Planning and Community Right to Know Act of 1986 (EPCRA), requiring states to create emergency response commissions and communities to form local emergency planning committees to prepare local emergency response plans for chemical accidents. It also requires chemical facilities to provide information necessary for emergency planning, and to submit annual inventory reports and information about hazardous chemicals. The statute also established the Toxics Release Inventory (TRI), which requires certain facilities to annually report the quantities of their emissions of toxic chemicals. These are for response not prevention.

The Occupational Safety and Health Administration (OSHA) was established in 1970 as part of the U.S. Department of Labor to assure safe and

Fundamentals of Industrial Chemistry: Pharmaceuticals, Polymers, and Business, First Edition. John A. Tyrell.

healthful working conditions by setting and enforcing standards and by providing training, outreach, education, and assistance. Subsequent to Bhopal, OSHA developed the Process Safety Management (PSM) standard which places accident prevention and emergency response requirements on facilities having listed hazardous chemicals above certain threshold quantities. Also with a focus on prevention, the EPA instituted the Risk Management Program.

Also, in response to the tragedy at Bhopal, the chemical industry voluntarily launched the Responsible Care® initiative. This began in Canada in 1985 but now applies in more than 50 countries with the major chemical companies adhering to the principles of Responsible Care®. Responsible Care® requires a focus on continuous improvement in the seven areas of Community Awareness and Emergency Response, Pollution Prevention, Process Safety, Distribution, Employee Health and Safety, Product Stewardship, and Security. Product stewardship covers not just the production and use of chemicals but also transport, storage, and disposal. Although this is a global initiative, there is widespread participation within the United States. Participation is a condition of membership for American Chemistry Council (ACC) member companies. The ACC reports in 2011 that the United States chemical industry had spent $10.8 billion in 2010 in environmental, health and safety programs, that releases had been reduced by 75% since 1988, and recordable injury and illness rates had been reduced by 74% since 1990 [1].

In 1998, the U.S. Chemical Safety Board (CSB), an independent organization established by Congress, began with the purpose of investigating accidents to determine the conditions and circumstances which led up to the event and to identify the cause or causes so that similar events might be prevented. They look for root causes of accidents and make recommendations. They also make their findings public and provide safety videos so that mistakes are not repeated. For example, in January 2010, in just a 33 hour time frame, there were a series of three separate incidents at a DuPont facility. There were releases of methyl chloride, oleum, and phosgene. The CSB did a thorough investigation of the root causes and issued a series of recommendations. A review of the incidences and the recommendations was put into a safety video so that others may learn from the accidents. The library of videos produced by the CSB on a variety of incidents is publicly available. Several companies show videos from the CSB collection at their safety meetings.

Sometimes it is not the "dangerous" chemicals that result in tragedy. I worked at one chemical site when they celebrated their 25th anniversary since the facility was built. This site had over one thousand workers and performed chemical operations from small laboratory scale to pilot plant and production. They produced chemicals in the order of hundreds of millions of pounds per year. Poisonous gases such as carbon monoxide, chlorine, and phosgene were used in large scale and many safety precautions were taken. In the 25-year history, there had been two workplace deaths – both from the same incident. The accident was not a fire or an explosion. It did not involve one of the poisonous gases. Rather it was from nitrogen. One worker went into an area that had been blanketed with nitrogen and was overcome due to lack of oxygen. A second person also died after going into the same area to rescue the overcome worker.

Common gases such as oxygen and nitrogen have been involved in tragedies in the NASA space program. In the early years of space exploration, NASA circulated oxygen rather than air in the spacecrafts. In 1967, while training for Apollo 1, there was an unexpected spark and three astronauts burned to death. In response, NASA began using nitrogen to blanket compartments prone to sparking. Years later, in 1981 five workers on the Columbia space shuttle entered a compartment that had been inerted with nitrogen. They hadn't realized that the compartment had not been subsequently flushed with breathable air. Three of the five died.

Unfortunately there are other cases of nitrogen asphyxiation. In November 2005, two contract workers performing maintenance at a Valero refinery died. One died while attempting to rescue the other. The Chemical Safety Board investigated this accident and issued a case study and safety video to help prevent more tragedies from nitrogen.

Chemicals have hazards and need to be handled with appropriate caution. However, they are often sensationalized in the media. Consider the next sentence which may be a cause for great alarm. There is one chemical that is the second leading cause of unintentional injury death for children aged 1 to 14 years, and the fifth leading cause for people of all ages. That sentence might cause the lay person to be terrified of chemicals. Can you guess what chemical it is? It is H_2O and the deaths are due to accidental drowning.

It would be a mistake to conclude that chemical tragedies only occur in industrial settings. Unfortunately there can be household accidents and accidents in academia. For example, in 2009 a researcher at UCLA died from injuries sustained in a fire that started when the plunger came out of the syringe she was using to transfer a solution of t-butyl lithium [2].

Despite the tragedies and probably because of the focus on safety, the chemical industry is safer than the lay person might imagine. OSHA maintains statistics on the number of reportable injuries and illnesses per year per 100 full-time workers. OSHA requires that a work-related injury or illness resulting in one of the following: death, days away from work, restricted work or transfer to another job, medical treatment beyond first aid, loss of consciousness, or diagnosis of a significant injury or illness by a physician or other licensed health care professional be reported. In 2010, the average number of reportables for all industries including state and local government workers was 3.8. Workers in chemical manufacturing and in resin manufacturing each had 2.4 reportables, about 35% lower than the average for all industries [3].

One reason for the relatively good safety record of the chemical industry is the focus on training and procedures. Each of the companies that I have worked for had an emphasis on safety. There were safety committees, safety inspections, mandatory safety training, and an overall safety culture. Many things are formalized. For example, new processes undergo a formalized Process Hazard Analysis (PHA) conducted by experienced people. Also, a Hazard and Operability Analysis (HAZOP) is done. Operating procedures are standardized (SOP) to prevent mishaps.

Inherently Safer Design (ISD) [4] incorporates several safety measures including the elimination or reduction of the hazard by using less hazardous materials or conditions or by using process or equipment design features that reduce either the likelihood or consequence of an incident. The risk is managed by using process control systems, safety instrumental systems, and mitigation systems such as sprinklers. Operating procedures, safety rules, training, emergency response procedures, and management systems are used to manage risk.

The concept of ISD is not new; it was introduced in the 1970s by Trevor Kletz [5]. One example is DuPont's response to the Bhopal tragedy [6]. At the time, DuPont was purchasing methyl isocyanate, the same chemical that killed so many in Bhopal, and shipping it over a thousand miles to a plant near a metropolitan center. That facility routinely stored several million pounds of methyl isocyanate on site. DuPont devised a new process based upon the ISD principles of minimization and substitution. They developed alternate chemistry that was based on another chemical which was converted *in-situ* to methyl isocyanate. This new approach reduced the amount of in-process inventory of methyl isocyanate to about ten to twenty pounds, a reduction of about five orders of magnitude.

Another example is an unloading system for hydrogen fluoride that was built by OxyChem at a Niagara Falls, NY facility. Hydrogen fluoride has a boiling point of 19.5 C. It is a severe pulmonary irritant. It is estimated that the lowest lethal concentration for a 5-minute human exposure to hydrogen fluoride is in the range of 50 to 250 ppm [7]. A 1987 release of HF [8] from the alkylation unit of a Marathon Oil refinery in Texas City resulted in the evacuation of about 4,000 residents and hospitalization of about 100. Fortunately, there were no fatalities. Another example that illustrates the dangers was an accident on September 27, 2012 when workers at Hube Global in South Korea were unloading hydrofluoric acid. There was a leak of HF. Five people died and more than 3,000 people were sickened [9]. In order to minimize the chances of a gas leak, OxyChem built an enclosed refrigerated building to house the rail car of HF. The car is cooled prior to unloading so that the HF would be in the liquid form. Any unintentional mishaps are then easier to deal with because the HF can be contained as a liquid.

Despite the efforts of the chemical industry, there have been lapses. In March 2005, there was an explosion and fire in the BP refinery in Texas City that killed 15 and injured more than 170 [10]. The accident occurred during startup of the unit and had several causal factors. Despite these tragic lapses, the trend is good; there is an ever increasing emphasis on safety and hazard minimization.

14.2 ENVIRONMENTAL COMPLIANCE

Under the Toxic Substances Control Act (TSCA) of 1976 (15 U.S.C. §2601 et seq.) and the Pollution Prevention Act (PPA) of 1990 (42 U.S.C. §13101 et seq.), the EPA evaluates new and existing chemicals and their risks, and finds ways to prevent or reduce pollution before it gets into the environment. TSCA governs the production, import, use, and disposal of certain chemicals. It is a lengthy law with 70 sections dealing with a variety of items including radon, asbestos, and lead paint. It requires testing of chemicals where risks or exposures of concern are found. Under TSCA, the EPA maintains an inventory of all chemicals that are manufactured or processed in the United States. One important aspect of TSCA is found in section five and is the requirement that notification be given to the EPA prior to manufacturing a new chemical substance. This is commonly called a premanufacturing notification (PMN). There are certain exceptions such as when a substance is manufactured in small quantities for research and development purposes and special procedural and recordkeeping requirements are met. Also in TSCA is a Significant New Use Rule (SNUR), which requires manufacturers to notify the EPA prior to engaging in the new use. The Pollution Prevention Act focuses on pollution prevention or reduction at source. The emphasis is on reduction rather than treatment of waste.

The Clean Air Act of 1970 was incorporated into the United States Code as Title 42, Chapter 85. It was amended in 1977 and again in 1990. Brevity is not the norm for government documents and this one is about 450 pages. This act regulates motor vehicle emissions and pollution from many sources, even noise pollution. Section 112 of the act is on hazardous air pollutants (HAP). The 1990 amendments included a program to control 189 pollutants including those previously regulated by the National Emission Standards for Hazardous Air Pollutants (NESHAP). The pollutants are listed in 42 U.S.C. §7412 and the list includes many organic chemicals. To give the reader an idea, the 14 chemicals beginning with the letter "A" are: acetaldehyde, acetamide, acetonitrile, acetophenone, 2-acetylaminofluorene, acrolein, acrylamide, acrylic acid, acrylonitrile, allyl chloride, 4-aminobiphenyl, aniline, o-anisidine, and asbestos. The law requires that the EPA Administrator periodically reviews and revises the list.

The EPA uses a technology-based approach to reduce emissions. Standards are developed for controlling emissions from each major type of facility within an industry group. The standards are based upon the emission levels that are already being achieved by the lower emitting sources within the industry. They are known as maximum achievable control technology (MACT) standards. MACT standards require sources to meet specific emissions limits that are based on emissions levels already being achieved by many similar sources in the country. MACT standards are a way the EPA has of promoting best practices. The EPA has issued more than 20 MACT standards for different air toxics. For example one of the MACT standards, published in 1994, reduces emission of 131 air toxics from chemical manufacturing processes. It requires reductions in toxic organic air pollutants emitted from places such as process vents, storage vessels, and other equipment. This MACT standard has a goal of a 90% reduction from the preregulated levels. Many of the MACT standards are not linked as closely to the chemical industry. For example, another standard, published in 1997 deals with incineration of medical waste.

In 1994, EPA promulgated (40 CFR Part 63) national emission standards for hazardous air pollutants (NESHAP) for the synthetic organic chemical manufacturing industry. This rule is commonly known as the hazardous organic NESHAP (HON) and established maximum achievable control technology standards to regulate the emissions of hazardous air pollutants from production processes that are located at major sources. This regulates emissions from five sources: process vents, transfer operations, storage vessels, equipment leaks, and finally wastewater streams, collections and treatment operations. Requirements for compliance, record keeping and reporting are given.

The 1976 Resource Conservation and Recovery Act (RCRA) (42 U.S.C. §6901 et seq.) involves the control of hazardous waste. RCRA governs not just chemicals but also municipal waste and landfills.

It was amended in 1984 to focus on waste minimization and to phase out hazardous waste disposal in landfills. In 1986 it was further amended to include regulation of underground storage facilities. The focus here was underground storage of gasoline. Recall that about that time, there had been some storage tanks that had leaked and a fuel additive, methyl t-butyl ether (MTBE) had been found in the groundwater as a result of leaking gasoline storage tanks.

The European Union regulates chemicals with a law called REACH. REACH is an acronym for the Registration, Evaluation, Authorization and

Restriction of Chemical substances. The law was enacted in 2007. Manufacturers and importers are required to gather information on the properties of their chemical substances and to register the information in a central database run by the European Chemicals Agency (ECHA). There is a general requirement for manufacturers and importers of substances of one metric ton per year or greater to submit a registration to ECHA. Technical information such as properties and guidance on safe use is required in the registration. For quantities of ten metric tons or more, a chemical safety report must be included. By using REACH, the European Chemicals Agency builds up a public database for both professionals and the public to use to find hazard information. There is also an emphasis on substituting for the more dangerous chemicals when suitable alternatives have been identified. Since inception in 2007, companies have registered more than 30,000 files describing the uses and properties of more than 7,500 chemical substances manufactured or placed on the market.

14.3 GREEN CHEMISTRY

Green chemistry has been defined as "the utilization of a set of principles that reduces or eliminates the use or generation of hazardous substances in the design, manufacture and application of chemical products" [11]. Twelve principles of green chemistry have been enumerated [12]: (1) It is better to prevent waste than to treat or clean up waste after it has been created; (2) Synthetic methods should be designed to maximize the incorporation of all materials used in the process into the final product; (3) Wherever practicable, synthetic methods should be designed to use and generate substances that possess little or no toxicity to human health and the environment; (4) Chemical products should be designed to effect their desired function while minimizing their toxicity; (5) The use of auxiliary substances (e.g., solvents, separation agents, etc.) should be made unnecessary wherever possible and innocuous when used; (6) Energy requirements of chemical processes should be recognized for their environmental and economic impacts and should be minimized. If possible, synthetic methods should be conducted at ambient temperature and pressure; (7) A raw material or feedstock should be renewable rather than depleting whenever technically and economically practicable; (8) Unnecessary derivatization (use of blocking groups, protection/ deprotection, temporary modification of physical/chemical processes) should be minimized or avoided if possible, because such steps require additional reagents and can generate waste; (9) Catalytic reagents (as selective as possible) are superior to stoichiometric reagents; (10) Chemical products should be designed so that at the end of their function they break down into innocuous degradation products and do not persist in the environment; (11) Analytical methodologies need to be further developed to allow for real-time, in-process monitoring and control prior to the formation of hazardous substances; and (12) Substances and the form of a substance used in a chemical process should be chosen to minimize the potential for chemical accidents, including releases, explosions, and fires.

The mnemonic, "PRODUCTIVELY" has been used [13] for the twelve rules. It stands for P (prevent wastes - principle 1); R (renewable materials - principle 7); O (omit derivatization steps - principle 8); D (degradable chemical products – principle 10); U (use safe synthetic methods – principle 3); C (catalytic reagents – principle 9); T (temperature, pressure ambient – principle 6); I (in-process monitoring – principle 11); V (very few auxiliary substances – principle 5); E (E-factor, maximize feed in product – principle 2); L (low toxicity of chemical products – principle 4); and Y (yes it's safe – principle 12).

Many, maybe even all, of these principles also make good business sense. Many of the principles have been employed prior to the concept of green chemistry. For example, consider principle one, "It is better to prevent waste than to treat or clean up waste after it has been created." This is true from a green chemistry perspective but it is also true from an economic perspective. It costs money to deal with waste. In the first place, you have purchased raw materials which are not being converted into saleable goods. Second, the waste has to be handled. It either has to be purified or money has to be paid for the disposal. Consider a simple example of a company that removes river water to be used in their process. After use, the waste water has to be disposed. It cannot be returned to the river unless it is first purified and tested to ensure that the water quality is demonstrated to be of sufficient cleanliness to be returned to the river without environmental impact. The purification and testing cost money. Equipment must be purchased to handle the waste water. If the process water usage is reduced by 50%, then less money is needed for equipment, handling, purification and testing.

Consider principle 11, "Analytical methodologies need to be further developed to allow for real-time, in-process monitoring and control prior to the formation of hazardous substances." In-process monitoring speeds the feedback loop and minimizes the time spent making out-of-specification material. For example, if a process sample is removed for testing and after a four-hour test, it is found that there is something wrong with the process, the manufacturer has been making out-of-spec material for four hours. An adjustment is made and another sample sent for testing. If the sample is still out-of-spec, inferior material has been made for another four hours. Without a fast feedback loop, as is achievable using in-process monitoring, inferior material is produced for several hours. For a plant running at a rate of 40,000 pounds per hour, this one experience can result in several hundred thousand pounds of material that is off-specification. It either needs to be sold at a discount resulting in a financial loss or reworked at an expense to improve the quality. In the worst case it must be disposed, which can be expensive. Clearly, the business priorities are in concert with this principle of green chemistry.

Like safety, green chemistry makes good business sense. The reader may wonder why all the fuss about green chemistry if it is something that might be done anyway. The answer is one of focus. Just as an increased focus on safety can lead to a decrease in accidents, the increased focus on green chemistry can lead to improvements both from an environmental perspective and from a business perspective.

In order to promote green chemistry, the U.S. Environmental Protection Agency sponsors the Presidential Green Chemistry Challenge. The awards are listed on the EPA website (http://www.epa.gov/greenchemistry/index.html). Some examples of past award winners are given here to illustrate the principles of green chemistry.

The 2012 Greener Reaction Conditions Award was won by Cytec Industries Inc. for the use of scale inhibitors for the heat exchangers used in the Bayer process that converts bauxite ore to alumina. Alumina is primarily used to make aluminum metal. Previously, the heat exchangers had to be taken off-line to remove the scale with sulfuric acid. Scale inhibition resulted in tremendous energy savings (green principle six) and in fewer cleaning cycles which means less hazardous waste generation (green principle one).

The 2012 Designing Greener Chemicals Award went to Buckman International Inc. for the use of enzymes to make paper of improved strength and quality. This enables higher usage of recycled paper, savings in the amount of wood consumed, and lower energy usage (green principle six). The enzymes are from natural renewable sources (green principle seven) and replace chemical additives (green principle five). The 2011 Greener Synthetic Pathways Award went to Genomatica for a microbe that makes 1,4-butanediol by fermentation of sugar (green principle seven).

Many organic reactions require solvation, typically in organic solvents. Organic solvents can create waste; some pose flammability concerns; and others may be toxic or harmful to the environment. For cost, environmental, and safety reasons, water is a preferred solvent but many reactants are insoluble in water precluding its use as a solvent. The 2011 Green Chemistry Academic Award was earned by Dr. Bruce Lipshutz of the University of California, Santa Barbara. He designed a surfactant that, in water, formed micelles which enabled many organic reactants to dissolve in the lipophilic inside region of the micelle. They reacted in these regions, effectively using water to replace organic solvents. This is a good example of green principle five: The use of auxiliary substances (e.g., solvents, separation agents, etc.) should be made unnecessary wherever possible and innocuous when used.

Because the potential for fires, explosions, and releases is minimized, this also adheres to green principle 12.

There are certain metrics used in Green Chemistry. One is associated with principle two: Synthetic methods should be designed to maximize the incorporation of all materials used in the process into the final product. That is the metric of atom economy [14]. Atom economy is the mass of the product divided by the sum of the mass of all the reactants. Ideally this number is one and the closer to one the better. Atom economy does not take into account solvent or other reagents such as bases. Another metric is the Environmental factor or E-factor [15]. The E-factor is the mass of the total waste divided by the mass of the product. Ideally the E-factor should be zero and the smaller the number, the better. However, some fine chemical or pharmaceutical syntheses have E-factors of about one hundred. Using micelles, the Lipshutz group reduced the E-factor of a Heck reaction from 137 to 7.5 and then even lower with recycle [16]. Consider the following experimental procedure [17].

> N-Methyl-2-nitro-N-phenylaniline (7c). To a solution of aniline 6c (0.182 g, 1.0 mmol) in DMF (5 mL) at rt was added freshly crushed KOH (0.252 g, 4.5 mmol). After 10 min, MeI (0.20 mL, 3 mmol) was added to the stirring mixture dropwise via syringe. Stirring was continued at rt until TLC showed consumption of the aniline starting material. The reaction was then quenched with 25 mL of deionized H_2O and extracted with 3 × 30 mL of EA. The organic layers were combined and dried over $MgSO_4$. The solvent was removed under reduced pressure and no further purification was needed to obtain the nor-halo N-methyl derivative 7c as a brown oil (0.194 g, 100% yield).

6c → 7c

We can calculate the atom economy by dividing the mass of the product by the sum of the mass of all the reactants. Three mmol of methyl iodide (molecular weight = 141.94g/mol) has a mass of 0.426 g so the equation becomes:

$$0.194/(0.182 + 0.426) = 0.32$$

We can also calculate the E-factor. The mass of the ingredients are as follows:

0.182 g aniline 6C
(5 mL)(0.944g/mL) = 4.72 g DMF

0.252 g KOH
0.426 g CH_3I
25 g H_2O
(90 mL)(0.902g/mL) = 81.2g ethyl acetate
1 g (assumed) magnesium sulfate

Adding these, we get 112.76 g. After subtracting the 0.194 g product, we see that the waste is 112.57. The E-factor = 112.57 / 0.194 = 580. This is a relatively high E-factor and if the synthesis were commercialized, undoubtedly modifications would be made to reduce waste.

The chemical industry has come a long way in the last five decades. There is an improved focus on safety, a greater understanding of the effect of chemicals on the environment, a stronger effort in waste minimization and a commitment to cooperate in areas concerning safety and the environment. Undoubtedly, as technology and understanding improves in future decades, we should see further improvements in safety and lessened environmental impacts.

QUESTIONS

1. Look up a past Presidential Green Chemistry Challenge Award winner. Give a one paragraph synopsis and list which of the 12 principles of green chemistry are exemplified.
2. Look up an experimental procedure in a current journal such as Journal of Organic Chemistry. Compare the procedure versus the 12 principles of green chemistry. Which principles are exemplified? Suggest possible areas of research for further green improvements. Calculate the atom economy and the E-factor.
3. Label as True or False: Based upon OSHA, the number of accidents and illnesses (OSHA reportables) for chemical manufacturing is only slightly higher (about 10% higher) than the average for all industries.
4. (Which best belongs in the blank?) In 1988, after Bhopal and other chemical disasters, the American Chemistry Council launched ______________ to respond to public concerns about the manufacture and use of chemicals. As elements of this effort, chemical companies assist each other to achieve optimum performance and report to the public.
   a. OSHA
   b. Responsible Care
   c. Good Neighbor

d. Better Living Through Chemistry

e. Chemistry is Safe (CIS)

f. The United Way

REFERENCES

1. Responsible Care® Fact Sheet American Chemistry Council July 2011.
2. Jyllian Kemsley. Chemical and Engineering News 2012; 90(5):10.
3. Bureau of Labor Statistics. U.S. Department of Labor, October 2011.
4. Dennis Hendershot. Chemical Engineering Progress 2012; 108(1):40.
5. Trevor Kletz. Chemistry and Industry May 6, 1978:287–292.
6. Victor H. Edwards, Jack Chosnek. Chemical Engineering Progress 2012; 108:48–52.
7. GJ Hathaway, NH Proctor, JP Hughes, ML Fischman. *Proctor and Hughes' Chemical Hazards of the Workplace*. 3rd ed. New York: Van Nostrand Reinhold; 1991. ,
8. John Atherton, Frederic Gil. *Incidents that Define Process Safety*. New York: John Wiley & Sons; 2008:101.
9. Jean-Francois Tremblay. Chemical and Engineering News 2012; 90(42):9.
10. Trevor Kletz. *What Went Wrong? Case Histories of Process Plant Disasters and How They Could Have Been Avoided.* 5th ed. Elsevier; 2009:414.
11. Paul Anastas, John Warner. *Green Chemistry: Theory and Practice*. Oxford University Press, Oxford, UK; 2000:11.
12. Paul Anastas, John Warner. *Green Chemistry: Theory and Practice*. Oxford University Press; 2000:30.
13. SY Tang, RA Bourne, RL Smith, M Poliakoff. Green Chem. 2008; 10:268–269.
14. Trost, BM. Science,1991; 254:1471–1477.
15. RA Sheldon. Chem. Ind. (London) 1992:903–906.
16. Stephen Ritter. Chemical and Engineering News 2013; 91(15): 23.
17. Andria M Panagopoulos, Doug Steinman, Alexandra Goncharenko, Kyle Geary, Carlene Schleisman, Elizabeth Spaargaren, Matthias Zeller, Daniel P Becker. J. Org. Chem. 2013, 78:3532–3540.

INDEX

Fundamentals of Industrial Chemistry: Pharmaceuticals, Polymers, and Business, First Edition. John A. Tyrell.

Printed in the United States
By Bookmasters